The Thinking Horse Breeder

A comprehensive guide to successful horse breeding

Jeanette Gower

Platypus Publishing

I dedicate this book to Peter Gower and my Mum and Dad,

who sadly never knew I was to publish this book.

***

Note:
For simplification purposes, the following text
will use "he," "him" and "his" to indicate all genders.

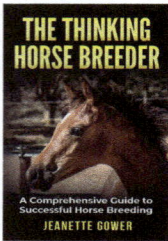

Cover photo by Janice de Gennaro

Australian Stock Horse filly, Chalani Regatta.

PLATYPUS
PUBLISHING

ISBN – Hard Cover: 978-1-959555-71-1

ISBN - Paperback: 978-1-959555-48-3

ISBN - Ebook : 978-1-959555-60-5

First edition published in Australia.

Acknowledgments

I would like to acknowledge the many people who assisted me with this publication; those in the industry who read my chapters to check for accuracy, editing and feedback. They have been brutally honest, with extremely helpful comments and valuable insights, which I have endeavoured to include in this book. They are listed in alphabetical order as all have contributed in a valuable way: Andrew McLean, Bev Schutz, Dr Kerry Mack, Dr Mel Simmonds, Frances Killmier, Janelle Groeneveld, Jenni Phillips, Karen Thrun, Katherine Szalay Evans, Kelsey Stafford, Kimberley McCallum, Lynda Rodgers, Merrie Elliott, Peter Haydon, Theresa Thomson-Jack, cover designer Amy Curran, and cousin Tim Ide, for his artwork. To my Substack readers, you have been motivating in no small way with your comments and enthusiasm.

I can't go past the people who provided the beautiful colour photographs which grace these pages. Without them, *The Thinking Horse Breeder* would not be the beautiful book it is, and would be far more expensive to you, the reader. I believe horses should be proven before breeding, so the photos reflect both this and our Australian heritage.

Lastly to my family for their support, encouragement and patience while I disappeared down the rabbit hole to write: Kim, Andy, Ashton, Xander, and Jason who was always asking "how is your book going?" *Below: Xander Ide (8yo) with his foal, Chalani Wizard.*

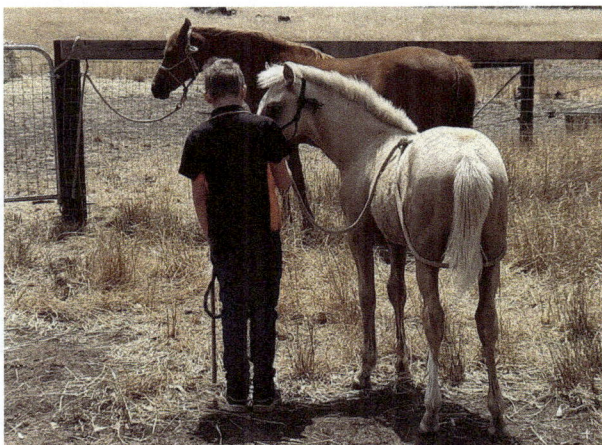

The 10 Commandments of Breeding

– Sheila Varian (Varian Arabians)

1	Breed toward your ideal horse. Don't be swayed by the voices of others. If you have done your homework, in time others will appreciate what you have accomplished.
2	Recognise what you are breeding and stay within your interpretation of the breed standards.
3	Breed equally for type, performance qualities, disposition, and trainability.
4	Never forget: No foot, no horse.
5	Always strive toward a horse of usable disposition, plus beauty. Neither is good without the other. However, if forced to make a choice, I'd keep the usable disposition.
6	Follow the lead your horses set for you. The next generation need not be similar in phenotype to the generation before, but each generation must be consistent in overall quality.
7	Breed forward. Look ahead. Wonderful new surprises may be awaiting you. Recognize them when they occur.
8	It is not difficult to improve the produce of a poor-quality mare in one generation. It is not even difficult to improve the produce of an average mare in one generation. What is difficult is to improve the produce of an exceptional mare generation after generation. That takes real skill, knowledge, gut instinct, and vision.
9	Don't be afraid to appreciate the qualities of other's horses. Breeding is a competition with yourself, not with others.
10	Consider your horse's attributes before you consider its negatives. All horses have both. It is for you to determine how positive his/her good qualities are before you dwell on the negatives.

Contents

Foreword

I can think of no one better qualified in this country than Jeanette Gower to produce a common-sense book, written in language understood by all, to cover every aspect on the subject of horse breeding.

In this year of 2023, the amount of information being disseminated on the breeding, raising, and educating of horses is truly staggering. Novice breeders can justifiably be completely confused, as can old stagers like me, as we try to absorb and work our way through it and grapple with technology as well. This overload runs the gamut from unrealistic theoretical notions from people of limited practical experience to writings presented in complicated language that few can comprehend. As would be expected, Jeanette has produced an immensely practical book written in a straightforward manner.

I first encountered Jeanette Gower in 1967 when she came to Willomurra while she was still a high school student. That was over 50 years ago. It is impossible to try to explain to anyone not around during that era just how backward we were with regards to horses, and the lack of information available to us. Any 'knowledge' about horses was loosely based on old English methods. While some inroads were being made by various dressage-based European horsemen who had migrated to this country after WW2, they were not widely accepted at that time. In SA we also had Tom Roberts, a man definitely ahead of his time in his understanding of a horse. I equate him to the effect Tom Dorrance had on the horsemen of USA. Maurice Wright was stepping out with demonstrations of the "Jeffery method" but in general it was not easy to find answers. A perusal of the *Hoofs and Horns* magazine from those years will show exactly what I mean.

To a large extent, many of us were ignorant of the fact that we were ignorant.

For some, it took the 1967 arrival of the Lougher family, and their highly trained Clover Leaf horses, to show us what could be done to improve ourselves and our horses. The burgeoning Quarter Horse industry also brought other Americans to this country, some who had a lot to offer - some certainly did not. An increasing number of Australians began to seek answers and experience overseas. While there had always been those Australian horsemen who could work their horses in a better manner, they were not into disseminating their knowledge; indeed, in most cases they lacked the language to try to explain what they were doing, and it is even doubtful that they knew themselves the rationale behind their methods. They just knew that whatever they were doing worked. Without doubt, it took the arrival of the Americans, those masters of the arts of promotion and communication, to provide a language of the whys and wherefores of horsemanship, especially as relating to the performance aspect.

Once we cottoned onto the fact that we were way behind the eight ball, the floodgates opened, and we lost no time in seeking knowledge wherever it could be found. I must say it was an incredibly exhilarating time and many of us felt we were on the cusp of something BIG, which was indeed the truth! Jeanette was right in the forefront of this quest for knowledge, a quest that has been ongoing for her for almost 60 years and shows no sign of abating.

From my earliest association with a very young Jeanette, it was obvious that she had been born with an objective and analytical mind. This, together with her communication skills, was further developed by the education she received along the way. From her own experience as a young struggling breeder with no land of her own, when every dollar had to do the work of three, she understood that, nice as it is to have top notch facilities, one can get by with pretty basic stuff – but you have to work harder, be smarter and have really co-operative horses.

Why do I believe Jeanette Gower is so qualified to write this book? Because she has been there and done it. She has done the hardest yards, enduring for almost 6 decades in a manner that no one in this image-ridden, fashion-conscious, promotion-minded horse industry of today, would have any hope of replicating. From two mares of obscure breeding, in whom only she could see any merit, Jeanette has, by thoughtful use of bloodlines, by keeping her horses in the public eye and having the fortitude to hang in there during low spots in the industry and through some massive knock-backs that would have encouraged more fragile-minded people to leave, she has successfully developed a line of Australian Stock Horses that are respected throughout the nation.

Jeanette is a true professional in the way she applies her knowledge, in her business deal-ings and her unflinching integrity. In an industry where there is increasing sentimentality and sooky notions and speech about horses, I have never known Jeanette to speak of foals as "furbabies" or call stallions 'boys' and mares 'girls,' let alone refer to herself as a 'horse mummy' (heaven forbid'). She understands that a horse is a horse and needs to be treated as such. She has an immense respect for them all and her professionalism is total.

Jeanette presents a number of excellent points for us to ponder. Although time and space will not allow me to comment on them all, I do believe that I must make mention of one of them. That concerns the matter of breeding for temperament as the main consideration. (Chapter 11.) Good dispositions are essential in any horse breeding program but bear in mind that a horse is supposed to be useful in some way. "Gentle" needs to be balanced by the need to keep a functional conformation and a trainable mind with at least a bit of "want to". If a line is never put to the test in some way, from a using perspective, a certain 'dumbing down' can creep in. Regrettably, this is already happening in some quarters. Not everyone wants to breed fast, switched on athletes with an overload of work ethic, but a horse breeder needs to keep some potential to be useful in his breeding stock.

I had a light bulb moment about this in 1975 in Texas. The great Matlock Rose presented a truly magnificent colt for our inspection. We were suitably impressed. Not so Matlock. He expressed his disgust: "Good lookin' sucker, isn't he, but the useless son of a bitch doesn't want to do *anything!*" I took his words on board.

So, we have this pragmatic lady who has done the hard yards, who understands that a horse is a horse and can be successfully produced and developed without constant "expert" in-trusion if principles of common sense and basic animal husbandry are brought into play. She knows that there is generally more than one way to achieve a desired result, depending on a person's individual situation. There are plenty of reasons why I believe Jeanette Gower is uniquely qualified to present this book. It should be mandatory reading for those already in the business as well as those planning to venture into the wonderfully creative, fascinating, and sometimes devastating world of horse breeding.

Merrie Elliott - Master Quarter Horse breeder, mentor and friend.

Bob n Stuff QH ridden by Kristy Banks, one of the many top performers bred by Merrie Elliott. Kristy is paralysed from the waist down as a result of a racetrack accident. You can see her leg strapped on in this photograph. She and Bob took out the High Point Qld Barrel horse for three consecutive years. Photo Jenny Solomon.

Work ethic and determination, Jokers Little Ace, QH, heading home. Another bred by Merrie Elliott. Rider is owner Amy Stemp. Photo Jenny Solomon.

Introduction

As early as I can remember, I loved horses, drew them, dreamed of them, was obsessed with them. I lived in the city without horsey parents. My mother always said, she couldn't get me to read, so she borrowed horse books from the library. I read every book available (and believe me, I read them all because there were so few in libraries back then), including classics like Hayes *Veterinary Notes for Horses*, and *Conformation of the Horse*. I knew every breed ever written up and turned every page in Encyclopedia Britannica with the word "horse" mentioned. I studied the racing pages. I had no horse friends, no money, and there were no courses or internet available.

Does this sound like you?

I asked for pocket money for all birthdays and Christmases to attend a few schools on borrowed horses. I became a stable hand at each year's Royal for well-known hackies, groomed at Adelaide polo for some years, joined Riding Club and the only local dressage club in the state. I listened in awe to Tom Roberts at his home every Friday night (while he was still riding and before he'd written his "Horse Control" books), fascinated by his Super-8 films on the re-education of problem horses. I worked on studs, including those of the famous Willomurra, gaining practical experience with youngsters, mares, and stallions.

All this happened before I ever owned a horse.

So when I sent a borrowed mare to a stallion to breed a foal, at 15 years old, with my own funds, I did it with some awareness of the horse industry and what I was doing. I must have done something right, as that bloodline is still in the stud!

The resulting foal, a filly from a "been there, done that" ex-polo mare Paradis, bred on Cordillo Downs station, was the first horse I ever owned. It engendered in me a huge desire to become a horse breeder. Believe it or not, I did as much research as was possible and had clear ideas to breed a horse with good conformation and talent. I was looking for an Anglo or Thoroughbred stallion. I rode the mare to the stallion as I wasn't old enough to drive and we didn't have a horse float. I left her there till she'd gone off, and rode her home again. She was in foal! The service cost the princely sum of $10. The stallion was later gelded and I was able to follow his successful hacking career. I was able to continue to lease the mare for another 8 foals. It also confirmed for me that I wanted to breed polo bloodlines.

Paradis with me riding, in the back yard, and at polo, with her owner,
Stan Grosser, Adelaide Polo Club grounds, first foundation mare for
Chalani, circa 1965

On reflection, I learned so much from this one mating, including a real desire to learn about the genetics of horse breeding, which was in its infancy. I wrote to libraries and researchers around the world by snail-mail asking what colour foal I might expect as I was very surprised to find that breeders couldn't tell me. I also broke in this filly based on techniques I had previously learned at dog training and the lessons of Tom Roberts, in spite of numerous people telling me not to do it myself.

Chalani Cat Ballou, second foundation mare
for Chalani, shown here as a yearling circa
1972. - Photo Peter Gower.

By age 19, I was invited to lecture at the SA Light Horse Breeders Association on the genetics of horse breeding and was lecturing in Equine Studies at TAFE in my early twenties. Over that time, I leased several horses, bought and broke in a few unhandled two year olds during my school holidays, and would sell each to pay for the next. I worked hard and asked lots of questions. There were so many old wives tales about, but that is another story.

Therefore, I have little sympathy for people who embark on breeding without interest in gaining knowledge other than through Facebook, and those who seemingly encourage it with "isn't it cute/gorgeous/stunning" comments. This indicates that they don't have a clue or are satisfied with mediocrity. It is not good for the horses.

Few people begin breeding these days with insight and awareness because it is so easy to jump in. Of course, they can love their horse, but that doesn't make it breeding material or the owner sufficiently knowledgeable to start horse breeding. Lazy breeders, ignorant breeders, and volume breeders don't listen to logic or accept help. They want shortcuts. They have no place in horse breeding.

After I married my husband Peter, we continued the stud, while calling the first meeting of the Australian Stock Horse Society in our state, only the second state to do so. We travelled Australia photographing and recording horses, attending Conferences and Expos both as speakers and horse exhibitors, and touring studs around the country. We watched demonstrations of the Jeffery Method by Maurice Wright, based on the teachings of that wonderful horseman Kel Jeffery.

Together we trained Sprint Racing Quarter Horses with our most successful homebred being Chalani Paper Tiger, who ran second in the 1980 All-Australian Futurity, (trained when in Canberra by Bill Hughes). I travelled the state classifying and judging ASHs, while Peter sat on the Board of the Society during the extensive planning for the opening ceremony of the 2000 Olympic Games in Sydney. I met some amazing people and had wonderful adventures at a very exciting time in the industry .

Early into our life came the wonderful polo sire, Rannock, a once in a lifetime buy, who was to put our stud on the map. It is said that you only get one great horse in your lifetime, so we were lucky to get him early in the career of our stud. When you know you have a great horse, you owe it to the horse not to waste that opportunity.

Rannock, "Wall of Fame" Australian Stock Horse, and foundation sire for Chalani, circa 1976. Photo Peter Gower.

Becoming a successful and responsible horse breeder is not for the faint-hearted. There are long days, late nights on foal watch, death of stock, injuries that need immediate attention, drought/fire/flood, and all manner of disappointments to overcome while trying to maintain family time and balance. Success is a hard road. It needs to be trodden with questions and worn through hard work before real understanding happens. Do you have the passion for the long road ahead?

They say "poverty is owning a horse" so why have 20 or so? Perhaps you've already bred a foal or two and you're beginning to realize how expensive and challenging it is. Perhaps your family bred horses for years, and now you are taking on some of the responsibility? How do you take it to the next level?

Yet with the birth of one special foal, it all suddenly seems worthwhile. The mare nickering softly to her newborn while it suckles, is always something to treasure, no matter how many times we see it. It can be a very long and arduous journey without someone to support you along the way and to guide you in the right direction towards mastery.

In this book, I am questioning long-held beliefs and aim to raise awareness. I know what it was like to be a beginner. Beginners need to start somewhere. But we do not want horses to suffer through silly mistakes that beginners might make. Beginners in horses have no place *breeding* horses. Breeding responsibly requires a much deeper understanding of multiple facets than simply horse ownership.

By turning these pages, you will be supported to follow your dreams and learn how to confidently make better decisions, save time, do better, and achieve partnership with your greatest love, the horses in your life. At times I have been provocative, deliberately challenging you to deeper thinking; at other times I have interwoven a little of my own story. If you are an established breeder you may feel the first few chapters are not for you, but work through them anyway. You will certainly get something out of them. Each chapter is complete in itself so you can skip to your favorite, or re-read for a deeper understanding.

Wishing you every success in your horse breeding journey – Jeanette Gower

Rannock - photo by Peter Gower.

Chapter 1

Successful horse breeding

Dare to dream

You watch a newborn foal trying to stand, with its mother softly nickering encouragement. You smile with pleasure at his fine lines, the quality oozing from its innocent eye, its proud bearing obvious from its first steps. This foal is that special one you have been waiting for, and you are in awe of this incredible moment. You are content knowing that you designed this match with the best stallion in his field and a mare you bred from your lines over several generations. This foal's future is bright, as you will be able to give him the best opportunity to thrive and do what he was bred for.

It gives you immeasurable pleasure to still marvel at mother nature and her gifts in the twilight of your career. That you still have a passion for breeding. This, you are sure, is what horse breeding is all about!

Imagine your future self. Take a notepad and consider where you will be in 50 years (or 30 years, depending on your timeline). Visualize yourself and take 10-15 minutes to write answers for these 12 questions. Be specific – for example, the answer to "regrets" below might be "I wish I'd gone to the World Championships," or "I wish I had visited renowned Irish Studs," or "I wish I had bred a foal by x stallion when I had the chance."

- How do you measure your life as a horse breeder?

- What have you achieved?

- What mark have you left on the horse world?

- What do you consider your successes?

- What is your reputation in the horse world?

- What risks did you take?

- What are your regrets?

- How many horses do you still have?

- What are you doing with them?

- Are you making wiser decisions?

- Are you content?

- Are you living the life you thought you would lead?

Now based on your answers, inverse your notes and thoughts. You are now a tadpole, starting out or starting again. Jot down notes on the next page. For example, if your answer to one of the questions above was "I wish I'd had more time...." Then use this book to find more time. Don't take 50 years to do it. Again, be specific.

If you skipped making notes, I suggest you go back before proceeding and continue taking notes. Save them. They will help to clarify and define your thoughts, and we will look at them at the end of this book. For example, one of my goals was to become financially literate so that I could earn passive income which would support my horses, while I lived off my wages. This I achieved by reading everything I could about financial success, putting money into Super, joining an investment club, and putting money into property and shares.

It is essential to visualize and understand where you want to be! The romantic image I have laid out above is unrealistic for most. Studs come and go. Most will only last 15 years. As a horse breeder, you need to know the *why*. My experience is that you must do it for love and not money. The faster you want success, or the bigger you want to be, the riskier it becomes, and the greater the likelihood you will fail or be forced to downscale.

- Why do you want to do this? How passionate are you? You won't succeed if you are wishy-washy or constantly questioning why. Are you resilient?

- What do you already know about horses? What is your background in horses? I suggest you won't succeed if you have no prior experience, such as riding, racing or driving.

- Are you emotionally up to the challenges? Are you prepared for what might happen to animals you love and the misfortunes of adversity, disappointment and death of animals?

- What is your financial situation? Can you afford to breed 1, 2, or 10 horses? How will you fund your horse breeding? Are you prepared to do the paperwork which comes with owning a business? Do you have property? How big an operation do you want to be?

- Are you keen to learn for the rest of your life? Do you want to be smart about it so you don't make common mistakes and can enjoy the journey? If you aren't prepared to study or gain practical experience, you will be forced to learn from the school of hard knocks. And believe me, they are hard!

- Are you prepared to follow the progress of horses you've bred over their lifetime? Over the years, I have bred nearly 300 foals and lost contact with only 20 of them once they left the stud. Following the horses lets you know if you are breeding the right horses which make your customers happy, and gives you promotional tools.

- Do you have the TIME required to put this into action? Time and exhaustion will be your worst enemies if you let them control you. Are you a person who can set up a routine, be flexible, and follow through? Do you blame others when you are exhausted?

- What is your self care? You need to look after yourself and be at your best if you are to look after your horses properly.

- Is your family supportive? Depending on your answer, how much help, financial or otherwise, can you expect? After all, it is your dream, not theirs.

- What are your values? Are you trustworthy? Do you have integrity? Are you reliable? Are you disciplined? Do you accept responsibility and accountability? How do you take criticism?

These are not idle words. On the contrary, all of these qualities are necessary to succeed in horse breeding.

"You can get motivated; you could attend a seminar as many do and get all fired up for a few days. But motivation wears off, discipline doesn't"
– Andrew Barnett

Chalani Aurora HSH, keystone mare, descendant of Paradis, circa 2014. Photo Kim Ide.

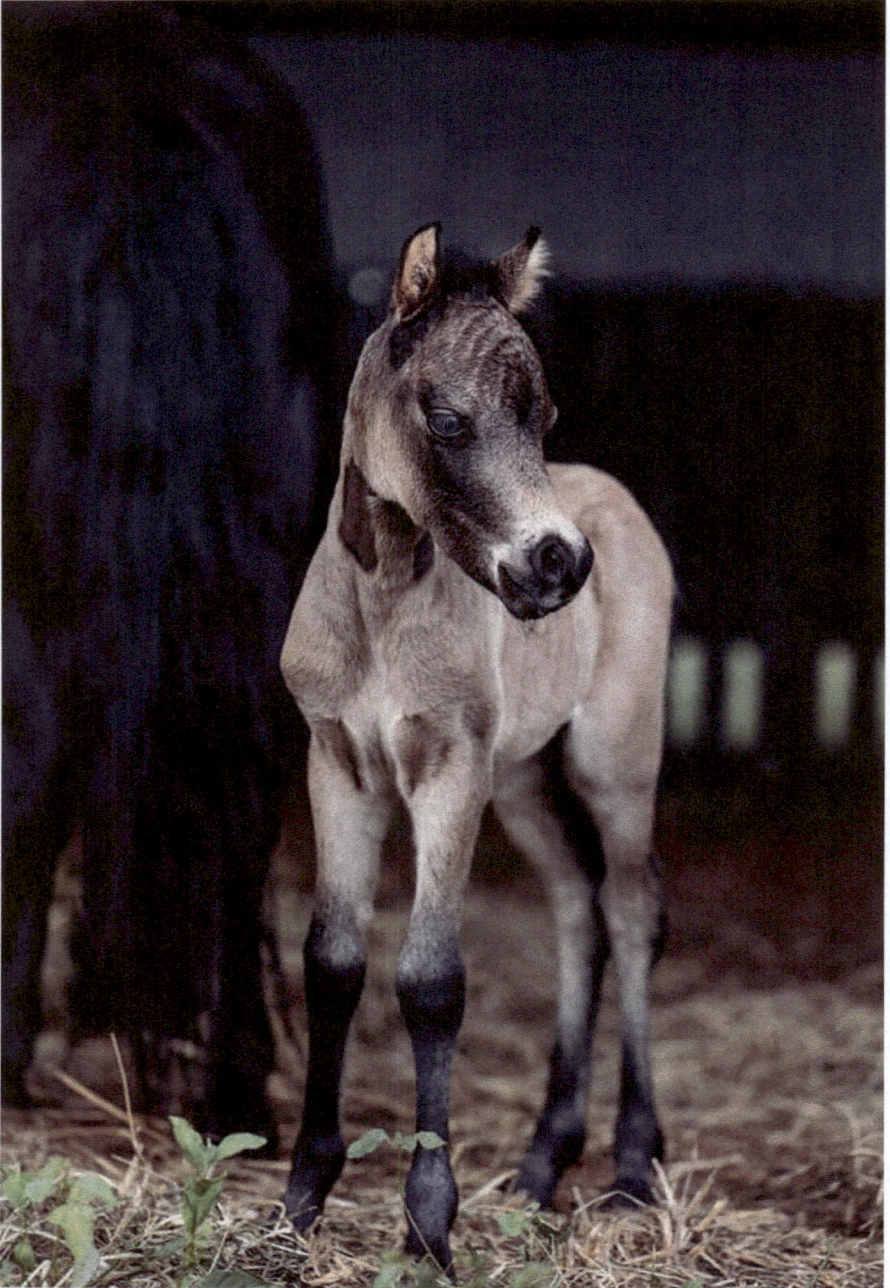

Shetland x Welsh foal. Owned / photo by Rebecca Dunstall.

Chapter 2

What is success?

Success is a staircase, not an elevator.

Success means different things to different people. So what does success look like? Write *your* vision of success in your notebook. It might mean:

- Respect of your peers.

- Number of sales or the sale prices achieved.

- Performance record of your horses.

- Performance record of horses you have sold.

- Satisfaction of buyers.

- Other breeders respect your lines.

- Your horses are bought as seed stock by emerging breeders.

- Your lines are absorbed into the fabric of the breed for generations to come.

- Your lines have become a new source of outcross for other breeders.

- Your work and genetics are not lost. It is there for future generations.

- You breed the type that is unavailable or difficult to purchase.

- People come to you for advice.

Why do you wish to breed?

Don't breed horses to make money, for status, for attention, for fame, glamour, or glory, to impress or because it seems "romantic." NEVER breed because the mare is "no good for anything else" or "it will quieten her down."

No one breeds "to better the breed." You can't improve a breed, only individuals. This cliché is used to justify breeding. Have breeders read and understood their breed standards? Can they quote from it? Do they know what *needs* to improve based on structural soundness and balance and an understanding of movement? Many breeders have limited knowledge of the rider's needs in a saddle horse because they haven't ridden much themselves! They do not understand the innate abilities of a chosen discipline or what is required to succeed in it. How, then, can they make responsible breeding choices?

Someone breeds those ill-bred, poorly-conformed horses, self-evident by the prices and wastage in the industry. I don't think you want to be one of those breeders!

Someone breeds these poorly conformed horses. But wait, all except the last are in foal!

I began breeding because I wanted a foal to reproduce the qualities of a favourite mare. I wanted to breed registered horses, so I chose the Quarter Horse breed. This morphed into sprinting Quarter Horses, but sprint racing never took off in Australia. So when we purchased our foundation stallion, the polo sire Rannock, we decided on a reset; to preserve Australian bloodlines, to continue with our Australian Stock Horses exclusively. Later it became Heritage Australian Stock Horses. Our reasons grew as our stud grew. But each reason was quite specific at the time and gave us direction.

I have my chosen breed because I love the breed, and I wish to breed to the highest possible standard. I breed because I love handling foals and raising youngsters. I love horses, and I love horse breeding - the art and the science, the lifestyle, and the friends who go with it.

So define your purpose, your goals, and your beliefs. Don't fall into the trap of doing what everyone else is doing. There is no value in FOMO (fear of missing out). You don't have to follow the latest fashion. Have confidence in your own ideas, and that all will eventually fall into place.

You will attract those who believe what you believe.

Now you are ready to write your mission statement. Every good business has one. It must be authentic to set you on the right path. Having a mission statement commits you to follow this path. It must be clear to others what you want to achieve. If you are a start-up or emerging breeder, this will likely change as you grow and become more specific. You may change your mission statement every few years as you refine your direction. But think it through now, based on what you wrote on your notepad. It needs to be something you can read daily and ask yourself if you are on track. It will describe succinctly and precisely what you are aiming to breed. Think of it as an "elevator pitch" or how you would describe your programme to a stranger you've just met.

Here is an example:

"Our goal is to breed horses that will be highly competitive, that have Warwick (campdraft) potential and will sell well at major sales. Toonga stud selects, above all else, a nice strong hipline and stifle position; not built too heavy, a fast horse keen to work a cow, with a good walk." – Des Castine.

Simple, explicit, and easy to understand.

Now construct *your* mission statement and post it on your website, bathroom mirror, or anywhere you can easily see it. This is not a New Year's Resolution to be forgotten after a few weeks. Instead, it needs to be something you believe in and something attainable. It will allow you to take the necessary steps to achieve your goals.

Remember, success is a journey, not a destination. It is a process of self-discovery.

Enjoy the journey.

Friesian mare Teske Van Berkley with her foal Utske Van Berkley. Owned / photo by Nadeen Davis.

Chapter 3

Code of a responsible breeder

Even though I wrote this code in 2009, it has withstood the test of time because it is always relevant. The closer you are to attaining this, the more successful you are likely to be. A Responsible Breeder (RHB) will:

<u>Breed for a purpose</u> – Have genuine short-term goals, a long-term vision, and a realistic plan to breed accordingly, that is, breed for a purpose.

<u>Start a breeding programme with quality mares</u> – Be prepared to remove (cull) those not producing your desired standard. There is hardly anything that a foal exhibits that isn't inherited. If you are a breeder, you will know this as a TRUTH.

Select breeding stock appropriately and responsibly. If you do not accept this truth, set yourself up to train horses, not breed them.

<u>Not use colour as his primary selection tool</u> – Colour must be considered if you set up a colour breed, say, Appaloosas or Paints. But a RHB will clearly distinguish between this and bad "breeding for colour" practices. Sadly, some people use colour as the only selection criteria and primary objective. If you own a stallion that 100% supplies that colour, you must be careful with the outside mares you accept.

<u>Set a realistic budget</u> – Before embarking on a breeding programme, research the cost and expenses involved in producing a single foal of your breed, multiplied by the number of foals you produce per year. Add in the overall expenses of property maintenance and the like, then compare with the average sale price you can expect to achieve for your breed.

Labour costs must be factored in, especially if you do not ride. Labour is time, and time is money. Decide how long it will take to achieve your goals and to break even (or if you ever will). If you are breeding multiple animals, you need to accept that it will cost money and time. It will significantly impact your family to breed and raise horses well.

<u>Prove his mares through performance</u> – Each breeding animal should be proven worthy of breeding by having PERFORMED (competition or useful work) before it is advanced to the breeding band. Having a uterus is no reason to breed from a mare. Unbroken

mares are unproven. Unless they are injured and carry a performance pedigree, unproven animals are a sure sign of irresponsibility.

Geld all but the very best – The breeder will make all decisions as to which colts bred are sire material. All responsible breeders geld before selling unless they have an exceptionally high opinion of a colt. As an RHB, you will be prepared to run on a good colt, if you think him to be that good, until the right person comes along. The breeder should make all gelding decisions.

Anyone who sells an un-gelded colt for someone else to geld is not a responsible breeder, whatever the age of the colt.

Produce better than the parents – Matings are planned wisely, not indiscriminately, haphazardly, or simply because of a trend. The breeder will know top quality and good sound conformation and will always strive to breed for this. He will be able to discuss conformation and bloodlines with others and the reasons for a mating. Each mating will be the best fit, complementing both sire and dam, to enhance the potential of the best possible result.

A TRUE breeder can breed better than the parents over each generation and that which he has bought in.

Be selective with outside mares and their quality – Check their genetic disease status - "I won't allow my QH stallion to be bred to any QH, Appy, or Paint mare that has not been tested for genetic diseases."
Be sure to check their temperaments. - "I will return a mare of doubtful temperament. If she has a foal by my stallion, it will likely get blamed for the resulting poor temperament in the foal."

Return outside mares in good condition – Care for outside mares appropriately and re-turn them in good or better condition. If extra feeding, farrier, and attention are required, the mare owner should be contacted and charged accordingly. This is DUTY OF CARE.

If unavoidable circumstances make this impossible, the RHB will communicate the issues clearly to the owner, provide options and negotiate the best possible plan. The stud owner will always strive to provide a responsible environment for the mare and an appropriate service regime, or not accept it in the first place.

Listen and learn from experienced, reputable breeders – Have an open mind. Take advice from those who have runs on the board (mentors, fellow breeders, buyers). Be prepared to listen, then add your intuition to determine decisions and goal setting. Volunteer your time to help a breeder you would like to learn from. Learn to handle horses of all ages well.

Become a horse person before becoming a breeder – Be a rider. Use and test your horses. If you have been a rider, you know the requirements and characteristics of a good riding horse because you know what a good horse "feels" like, and you know its quirks. You either ride or listen to those who ride your horses, especially regarding the good or bad qualities they are likely to reproduce.

If you are not a rider, you will have to pay someone to do this work for you, or your stud will rely on selling unbroken youngsters, an already oversupplied market.

Do research, research, and more research – Research management, nutrition, health, veterinary, genetics, pedigrees, and history. Find leading breeders and reputable breeders (not always the same, but if you do your research, you will work that out). Take courses, and read good books on breeding and management. Theory cannot replace experience, but without theory, practice is clueless. Finally, never quit researching and stay up to date. This is a lifelong task.

Research the market and breed within the market – Be alert to market changes, particularly in times of oversupply or drought. Be prepared to pull back when the market flags, and do not dump horses regardless of suitable homes. Matching the horse to the buyer is essential. A responsible breeder will continually assess the market and not over-produce or mass-produce. Volume breeding is just flooding the market. True commercial breeders who can supply a definite market and make a go of it, are few and far between.

Represent horses in a good light – Responsible advertising and marketing means representing a horse truthfully and disclosing fairly regarding suitability for purpose. Consumer laws and duty of care require this as a minimum. Disclosure includes providing suitable history (worming, inoculations, etc.), degree of handling, genetic disorders, and tests, if any, and similarly of the parents. This is to find the best home that fits the purchaser's experience.

Unethical practices and/or ignorance frequently mean that horses are advertised in a poor light for height, colour, and age. It can mean the drugging of unsound or temperamental horses, bad photos, and even breed and pedigree may go out the window. Buyer beware; as by definition, a responsible breeder will not do this.

Not ignore traits that negatively impact the horse – This refers to conformation, soundness, and/or health. Don't be influenced by "extreme fads." Fashion changes, and you could be left on the back foot. This includes crossing carrier to carrier of tested genetic conditions. Anything impacting a horse's performance negatively is a bad breeding practice.

Not criticize or gossip - about other breeders, judges or competitors, or their horses. Be conscious and aware of the "frenemies" - who will watch your progress, comment on social media, and then not speak with you face to face as they cannot mask their intentions. They will seek opportunities to mislead or make derogatory comments. The market knows who they are. They will likely be out of business in 5-10 years or trying another breed.

If different judges don't place your horse, seek to understand what they are seeing rather than falling into criticism of the judges. You will be tarred with the same brush if you stoop to this level.

Match the horse to the buyer – Keep every home-bred horse until you find a suitable match to the best of your ability. Do not simply sell it to the first home. This amounts to "don't breed more than you can afford to keep," as a sale is not guaranteed when you want or need it. For example, an injury can occur.

If you breed 10 foals a year, you usually need 10 sales yearly to keep your numbers steady. That is a lot of buyers to find in one year, each year, on top of your other stud duties and workload. Not everything you breed will be what you want or need to keep. You are now in the business of selling! You have become a salesperson. Is this what you really want to do?

Breed only what he can work with – Make sure any youngster you sell knows "the handling basics" before leaving the stud. This is one of the best ways to ensure your horses don't end up at the auction yard down the road. A few known studs breed huge numbers of foals a year, pick out the best half dozen, and send the rest to auction yards without handling or papers. These are NOT responsible practices. Avoid buying stock from such places.

Understand that horses that go for slaughter reflect a failure of equine breeding principles.

Give horses the best possible handling and start in life, reducing the possibility of that fate. Breeding for quality and for the market is the best protection against future abuse or neglect. For the aging and injured, euthanasia is the only ethical option - not dumping at auction or being sent to slaughter.

Have the paperwork up to date and available for viewing – Prospective clients require appropriate contracts drawn up, fees paid with respective societies, information on regulation criteria, and contact details for further information. A RHB will not mislead regarding paperwork, service certificate availability, DNA tests and results, branding/micro-chipping, or signing over documents. Neither will he unduly delay its processing. Anything over 30 days is considered an unreasonable delay to be avoided.

Offer follow-up information to buyers – Have faith that your product is suitable (because that's what you've sold to them, right?) If a buyer has trouble, be prepared to assist through advice or in person. Things that are unforeseeable or out of your control happen, but if you can help, it is good for the horse and public relations. If you do all the above, there will not be a need to feel guilty or disheartened, and such issues will be rare.

Offer to buy back or re-home horses bred. If it no longer works out for owners, rather than have sold horses go through several homes, offer to buy back if possible, or assist with the sale.

Be a member "in good standing" with his respective Breed Association or discipline. Have a history of active involvement as a competitor, organizer, volunteer, sponsor, or administrator, and be known as a supporter of your industry.

Have a handover plan – If you can no longer care for your horses due to unplanned ill health, accident, or death, ensure your relatives know what you want done with them. You may wish to redo your will to include instructions for your animals. Get income protection insurance, and don't rely on your partner to fund your hobby. People split up!

If your relatives are not horsey, make sure they have the names/addresses/people who can help and assist. If they suddenly have to care for your horses without your input, your horses will most likely be sold off to the first available buyer or sent off on a truck.

Understand that he is in business – Conduct yourself with appropriate professional and ethical standards. Being in business means breeders are subject to critique, and their actions reflect the industry. Responsible breeders understand this and practice all the above, knowing that not only are they representing themselves but their breed, their association, and all breeders alike.

After reading this, you might think, WOW – I didn't know so much was involved! Good. It is desirable not to jump right in without an action plan. How much time will you require to implement this? What routines will be required around your current job or workload?

Revise your mission statement and your dreams in light of the above. How do you see the code applying to you? How will you put this into practice?

Rachem Isabelle Australian Pony, and foal Kyabra Park Devil Wears Prada. Photo by Tegan McKenzie.

If you would like a copy of this Code to print for future reference, feel free to email me with the words CODE in the subject line. I will be happy to send it to you. My email address is at the back of this book.

Ironhorse Behold The Dream. Owned / photo by Iron Horse Appaloosa stud.

Chapter 4

Think before you breed

One of the first decisions you will need to make is the size of your operation. Will you be a hobby breeder, breeding perhaps one or two foals every few years, a small stud (up to 5 broodmares), a larger stud, or a commercial breeder, like most thoroughbred studs?

Choose what will work for *you*. A lot depends on your available land and how you will operate. It is said that if you raise horses like cattle, you must manage them like cattle. Your facilities must be set up for large numbers of unhandled horses, and you must plan how your facilities and staff will manage them. The other extreme is intensive farming on very small acreages or large commercial thoroughbred studs.

All of these systems can ensure responsible breeding but require very different management styles. The boutique breeder tends to think that it is irresponsible if you are not there for foaling. The range-land breeder sees the horses in their natural environment and cannot understand why someone would spend thousands of dollars saving an 18-year empty mare with $30,000 worth of colic surgery. The boutique breeder sees a barbed wire fence and is horrified, whereas the primary producer has no problem with it.

Each operation on a large scale will require staff to run it. You may be a considerable distance away and maybe not have a lot of input into its running. Its success will depend on the appointment of the right management. On the other hand, a small stud can be run by one passionate person who knows what they are doing and follows the basic principles in this book.

Family support is usually essential. Very few people can run a stud alone. So how do you do this if you have a disinterested partner? You know, the one who is not into horses, doesn't like an outdoor lifestyle, likes animals but not the work involved, doesn't like farming, or prefers a city lifestyle?

You really must have a sit-down conversation about this, especially if horse breeding comes after marriage. You must decide if you will share finances and what things you will do together. What will each other's roles and responsibilities be? Have regular date nights and support the other partner's interests. Don't talk "only" horses. This is a definite marriage breaker. Become an "interesting" person, even if it is merely watching the news together, so you have mutual topics of conversation. Who will do the cooking and other basics while the other is outside? And if you are a woman, what tasks will your partner take on willingly if you are pregnant?

Who is your mentor? Everyone needs a mentor, or perhaps several. A mentor is the first person you can turn to for advice, someone you respect and trust to keep information to themselves. If so, you must reciprocate that confidentiality. They need to be able to trust you. A mentor is someone who is skilled in the knowledge you would like to acquire and doesn't mind sharing it with you. They can usually assist you with networking. Most people like to share their knowledge with someone enthusiastic, so make sure they enjoy the conversations too, or they will feel "used". Mentor wisdom is precious, so take care not to lose it by some senseless act.

Your practical skills and knowledge need constant updating. What do you need right now? Take courses to fast-track your understanding. Your mentor will usually suggest things and give tips and tricks when a problem arises. You may have to pay for this help.

One of the essential things your mentor can help with is to open your eyes to what can go wrong, so you are aware of the bumps in the road. Your goal is to become a successful, *thinking* horse breeder!

Importantly, if you buy something, ask your mentor first what he has to sell. Your mentor will usually get nothing out of this relationship with you except the enjoyment of seeing you succeed. In return, the best thing you can do is repay him by becoming one of his clients!

What is your emotional connection with horses? Running a stud yourself or as a family relies on your emotional connection with horses. You need to really enjoy working with them. If, at any point, it becomes merely an obligation, stop breeding and reassess where you are. It must be fulfilling to be around them. There is nothing wrong with taking time out from breeding for a year or two. Believe me, the time will pass quickly.

Understanding facts and figures, having academic knowledge, romantic aspirations, and the love of horses is not enough without understanding the horse's nature and behaviour, and being able to read this at every level when you are with them.

This transcends every other skill you may have because, without it, you are not a horseman, merely a horse owner.

Realizing what can go wrong

Most people breed their first foal because it will be "fun," and nothing goes wrong. Or you might lose a mare and never breed again.

There are so many things that can go wrong. Your horse severely cuts itself on Christmas Day just as you are getting ready for family dinner. You are about to attend an important funeral, and your mare starts to foal. You are about to travel overseas for work, and your horse is suffering from colic. You're with the vet, and the semen missed its pickup from the terminal some four hours drive away.

Can you expect the unexpected? Are you flexible? What arrangements can you make with short notice to cover contingencies? Are your neighbours available to assist with picking up children, for example? Having contingencies in place will allow you not to become anxious and to think calmly when making decisions. Don't overthink things, as it will only cause you more distress. Do you become emotional or procrastinate around decisions? Teach yourself not to get worked up. That way, you can remain calm no matter what happens.

As you become more experienced, your calmness will allow you to draw from your emotional and practical toolbox to problem-solve efficiently with less distress, not only for yourself but also for those around you.

Integrity

Integrity means being honest and having strong moral principles. You demonstrate it by your actions with people, clients, and staff. But it is more than that. It is being honest with yourself and accountable. If something goes wrong, and it is because of something you said, or should have said, or something you did, or should have done, *own* it. It is ok to say, "I am very sorry. I never thought about it that way." Having an honest discussion brings people together to understand differences. If you promised something and didn't follow through, *own* it. "I am very sorry; I will send the paperwork tomorrow." DO IT! No excuses!

When my parents were arguing, it was always about who was "right" or who spoke the loudest and which one gave up first. When I went to university, an argument was a "debate." Anything was up for discussion. We debated any viewpoint as long as it was a logical progression of the argument, not a deflection or an attack. Let yourself understand the difference between a rational debate and a senseless argument. Bowing out (for example, on social media) will prevent you from stooping down to that level.

Your horses also demonstrate your values. Horses don't lie. They think and respond to how they perceive you. They are authentic. It is easy to see how horses relate to their owner the minute you visit a stud. Equally, visitors can see by your horses' reactions how they have been handled, and if it has been with calmness, care and kindness.

Your horses are a reflection of you.

Additional responsibilities of Pony Breeders

It may seem unnecessary, but I will say it anyway. Pony breeders are responsible for catering to children and their parents by producing ponies with a suitable temperament. They need someone competent to break them in (a family of competent children?) and the ability to communicate with parents who so often are not horse people themselves. This adds responsibility and accountability (and frequently worry) after the sale. Have reputable people lined up for recommendations and information to share, above and beyond your own expertise. You may need to educate newcomers about the perils of laminitis. Children outgrow ponies fairly quickly. This means you compete with re-sellers for sales and marketing.

The last call

Unfortunately, where there is life, there is death, and sooner or later you need to make that last call. Can you make the necessary decision without further suffering? An RHB will responsibly retire breeding stock or arrange euthanasia to prevent them from undergoing neglect or abuse or reaching the slaughterhouse. Any decision must cater to the animal's needs, not the owner's emotional needs. Can you do it? If not, don't breed horses.

Think before you breed.

Lynx Little Commando, QH. Owned / Photo by Jodie Metcalf

Chapter 5

Starting out or starting over

What is your breed? Breed selection is a personal choice. Is your breed relatively rare, barely known, or popular, as this will affect your promotion and how you operate. We will go into this in future chapters.

You have to decide which disciplines you are targeting and if you wish to specialize in one discipline or breed horses that fulfill the requirements for multiple disciplines. These decisions will affect your choices of breeding stock. Do not be misled into thinking that breeding crossbreds which cannot be registered is a better way to go. It means you are cutting out much of your potential market.

Know what is "special" about your breed or type and the requirements of the discipline you enjoy. What is unique about your breed? In addition, you need to know the standards, rules, and people in the industry, not only the horses.

One of the greatest mistakes of an emerging breeder is to choose the wrong breed, then try to turn it into something it is not. An example might be that you like Scottish Shetland ponies to be fine, and the breed standard says they are intended to be stocky. Don't try to change the breed. Don't ruin a breed by attempting to reinvent the wheel. A similar thing happens when breeders cross inappropriate breeds. Instead, choose a breed that is already designed for the purpose. Then, you are less likely to fail.

Peas in a pod. Skyeblu Paint Stud. Owned / photo by Skye Burke.

Define the type of horse you want. Your type should be immediately evident when someone visits. There should be no need to describe it. If your horses are like peas in a pod, all the better. People then know what you are offering and what their qualities are.

Finally, in this age of breed specialization, I suggest that it is not enough to breed "the athletic horse" or the "show ring" horse. The athletic horse must show quality, and the show horse must show athleticism. In addition, it is not enough to produce an athlete that only a professional can train or a show horse that you can't ride anywhere outside a show ring.

Some breeders say they breed for temperament, not looks. This statement is a cop-out and a sign of mediocrity. Breeding for temperament does not mean that looks (structure and soundness) or ability (trainability and performance) go out the window. The type of temperament you want should be guided by what you like best, enjoy getting along with, something you don't need to make excuses for, and what you believe your clients require.

Horse breeding has moved forward in so many ways in the past century. I look back at many old lines and think how horrible some "popular" ones were, but there was still a reason they were "popular." I also look back and see some remarkable horses. I read their histories and marvel at how good they were, their extraordinary feats, and the philosophy of those breeders earning a living by breeding, competing, or simply *doing* things to prove their horses in work. Unfortunately, we are rapidly losing the functional ability within breeds and genetic diversity, thus creating a gridlock that is difficult to reverse. It is up to responsible breeders to ensure we do justice to our breed for it to survive into the future, based on the breed having a *purpose*.

Initially, I chose the polo pony type because I saw these as tried and tested for soundness, temperament, and athletic ability, around 15.2h and a ready market. In addition, I wanted a good head of clean lines and a great eye. After all, this is what you see every time you walk out to your horses. It also indicates quality. I want horses that are free moving, light and comfortable to ride, with a good work ethic. Choose the breed that gives you the qualities *you* want.

The important thing is to take your time and stick to your goals. Start small to grow later.

Overnight successes are rare and are hard to duplicate.

Breeding for colour

If it is a good horse, colour doesn't matter. However, if a buyer only wants a specific colour, and you don't have it, don't think of it as a lost sale. There will be someone else who wants your colour or doesn't care. A buyer seeking a purpose and performance will not care about its colour, or otherwise will pay a premium for the one that fits both colour AND purpose.

If you breed for colour, you have additional responsibilities. First, you must know the various colours' genetic makeup and how your colour is produced. You must have your horses colour tested to plan matings and predict results. You need to understand this well enough to explain it to clients. The companies which test for colour explain everything on their websites.

Stallion owners have the responsibility to filter outside mares. For example, you must advise mare owners who have an overo, that you cannot accept an overo mare to your overo stallion. (This should be mandatory with all associations where colour is a criterion for registration.) Or if a dilute, and you have a dilute stallion, is the mare owner aware they might end up with a double-dilute foal?

Mare owners might choose homozygosity in a stallion if they don't want a chestnut foal. Some chase white markings (bling) as these are "showy" but consider if your horses are in country where they are easily sunburnt, or prone to cracked heels. Pink skin is prone to cancer of the eyes or genitals. You won't want double dilutes, white faces or socks, if you can't give them the attention they need.

Never breed for colour because you think you will get a higher price, or it will sell itself because of its colour or "rarity." Believe me, others think the same thing, and they are not rare for long. Having seen many bad horses of "colour" we decided to breed a few, believing we could produce better-pedigreed types than what is out there. You might think the same, so very quickly, these colours become mainstream.

Joseph's Dream Appaloosa stud Namibia. Photo Joseph's Dream.

Naming and Personal Branding

Choosing an appropriate name for your stud will have a significant influence down the track. The simpler, the better. It may be an existing property name or have familial/historical significance. It may be a mix of names of the principal owners. It doesn't matter. It should be short enough for a driveway sign and on advertising so that the name is "out there." Can it be easily pronounced? (Test it on your friends). Is it distinct, and can it be remembered easily? You don't want it to be confused with others that sound similar.

If a property name is too long, try to shorten it. Is it a name that can stand alone? For example, Willomurra Stud was only ever known as "Willomurra."

Before settling on a name, check with your accountant if you need to register it as a business name and what name your business will be in. This may affect what name you register horses under and if you can be considered a primary producer, with its accompanying tax breaks. Will you use a prefix? If so, you must register your prefix with the Breed Society. It will usually be the same name as your stud, or a shortened version of it, so consider both carefully. Too long a prefix and it becomes limiting for selecting your horse's registered name. A good prefix is invaluable for your stud promotion, as it indicates to all who the breeder is. It is perfect advertising in any place with an announcer. Hence it needs to be easily pronounced! The repetition means it becomes associated with your horses.

You will want to have a brand. In the current climate of microchipping, I still see a need for brands. It is one way a horse can be traced, especially at an auction where a horse's brand can be seen from a distance. This has frequently allowed a lost horse to be traced or even rescued. For this reason, we will freeze brand at weaning time. It leaves a white brand on the shoulder of the horse. (Breed association requirements decide the actual position). We record all our brands by year in a brand book and on HorseRecords.

I recommend designing a neat, simple brand suitable for youngsters who grow. There is no need to brand so you see it by helicopter! It is best if you can incorporate a symbol or logo in it. I always remember Greg Lougher with his Clover Leaf stud, Clover prefix, and a Clover Leaf as his brand. Clever. In Australia each state has different regulations on registration of brands, so check what you are allowed to use.

When you name your horses, having a system in place is best. Let it reflect the general tone of the breed. For example, using Docs Cracker Jack for an Arabian, or Star Princess of Saudi, for a Quarter Horse would be a mistake. Find something catchy or appealing. Calling it Bones may be funny at the time, but I believe horses end up living up to their name. A horse with a silly or unbecoming name rarely becomes a household name. Just like it is thought that owners end up looking like their dogs, I believe a horse's name reflects the breeder and the horse.

Our strategy is to have a theme for each mare. All of Cat Ballou's foals were named after films. Her dam was National Velvet. All of Paper Tiger's foals (a daughter of Cat Ballou) were named Paper something. Once you choose a theme, it is easy to expand on it as you breed more horses. It is easy to track the mare lines back in a pedigree. We use the stud name "Chalani" (Aboriginal for place of cockatoos) as the prefix before each name. If we can include something of the sire's name at the time, we like to do that too, such as Chalani Mirage, who was by Chalani Mystic out of Short Sighted (TB).

Others select names alphabetically, "A" being for the first year of breeding. There is also a free Names Generator in HorseRecords. (see chapter 32). We keep a list of names, built up over many years, to select from once the foal is born. I see breeders who can't settle on a name for months and, therefore, cannot register them. For the buyer, this can be a disincentive, especially if the horse, in the meantime, is called something dubious as a pet name.

Make sure the name is a name, not a sentence! Too long a name, too airy-fairy a name, something in another language that no one can pronounce, or a name that is hard to read are all avoidable with a little forethought. Also, always think of the buyer when you name a horse. No one wants to buy a horse with an awkward-sounding name.

Horses live up to their names.

Chapter 6

Financial mindfulness

The only way to make a million dollars in horses is to start with 10 million.

It may be tempting to skip this chapter, but it is important. How you look at your budgeting and financials will be one of the most determining factors in your success. Do you want to know how you are progressing and thrive in the future?

To be in business, and that's what you are since you are buying and selling, you probably don't need a business plan, but you do need a *strategy*. So that's why clear goals are essential. Being a successful breeder takes consistently producing great quality stock and more than one generation on the ground. It's an initial 15–20-year journey of time, effort, energy, and $$ to do it.

Anyone can get a mare pregnant, land a foal, and perhaps have it broken in as a 3-4-year-old. I suspect this last bit is where a lot of people become stuck and the slow road downhill begins. They don't handle their youngsters well enough and have difficulty taking the young horse to the next level. Therefore, a flood of 18mo-4yo unbroken and green-broken horses is on the market.

Even worse, is the person who has a young, unbroken mare they have put in foal to sell, because they think that will add value. Nothing could be further from the truth, as you have narrowed your market to the non-riding sector, and lost any breeder who doesn't know or like the stallion she is bred to.

Let's face it, young riding horses are work, and socialising them and taking them out can be tiring or confronting. That's in addition to the others you are breaking in or handling. You then realise how hard it is to raise youngsters and ride them as 3 and 4-year-olds, costing you thousands more than you make for the privilege. Much more than the $$ you make successfully selling the occasional foal as a weanling.

Only the well-established studs manage to sell all their youngsters as weanlings or yearlings. Breaking in at the time of writing is around $A3,000+, and the wait to break a horse in can be six months or longer. Others pay someone to do the riding "on stud." This is one of the major costs of running a stud. Another is the routine and emergency veterinary costs. These can rapidly head out of control! Then there is horse feed, travel, and the like. Take the time to appraise your annual, regular, and unexpected once-offs from the past years to get a picture of your running costs. If you don't know what they are, it's time to work it out.

A very good breeder once said to me: *"Horse breeding is extremely expensive.........*

...... for my husband!"

Horse income cannot be relied upon from year to year. It is impossible to enter horse breeding to "make money." This is not to say it is impossible down the track, but not until you have reached the summit in reputation and spent years at it. Breaking even is the best you can hope for in the first generation.

When the market goes pear-shaped

Having been involved since 1967, I have seen a lot of market cycles in the horse industry. Anything with a uterus was bred when it was booming, and the busts which followed saw many not continuing.

I have seen the warning signs of oversupply. I have seen people come and go, good studs drop out, and young upstarts disappear. But, on the other hand, some breeders have continued because they started with a strong foundation of ethics and integrity and knew what they wanted to produce and adjusted for the market. (Note - adjustment doesn't mean according to fashion, it means according to demand).

Cycles are dependent on many things: drought, flood, prize money, sponsorship, the backing of a strong association, usefulness of the horses, disease (Covid being an example, though there have been many, horse or human), fashion, new breeds entering the market, and the economy. But the one law that applies to all is SUPPLY and DEMAND.

You can spend thousands of $$$ more than someone else to produce a horse to a saleable stage, but it still won't be worth more than what the market will pay for it. This can be a serious wake-up call for some breeders.

There has always been an over-supply of the wrong types of horses and an under-supply of good quality educated stock. There has always been an over-supply of unhandled, unsound, or poor-quality horses coming onto the market. But when the market is about to slow down, warning signs are always there for those doing the right thing. For those who are conscientious enough to produce what they can handle and sell to good homes, their horses will always sell, though not necessarily for the prices they'd hoped, nor in a timely manner.

Good horses will always sell, no matter what.

Some of the warning signs of a downturn in the market are:

- Cattle prices are going down.

- Higher than usual price of feed.

- A larger than usual number of stallions coming onto the market.

- People dropping prices on good quality stock.

- Good quality taking much longer to sell.

- Horses being sold for under the stud fee.

- Average horses or horses with limited potential practically being given away.

- Stud reduction sales increasing.

- Huge numbers of adverts for free, or "must sell" horses.

- Sellers "changing direction."

- The prices of non-mainstream breeds, once quite expensive to obtain, dropping to the price of, or below mainstream horses.

- Stallion owners dropping service fees.

- Organisers finding it difficult to get support and entries for shows.

- Limited demand for good types of very useful geldings.

- Increasing numbers coming through saleyards and to slaughter.

- Studs with larger numbers of young horses yet to sell.

- More owners and breeders not giving their horses the care and attention to health and nutrition than usual due to market conditions, drought or other factors.

- Negative reports from vets and others in the industry in the know.

- Slowdown in retail trade and associated horse industries. (Horse goods are considered "luxury" items.)

- Horse properties coming onto the market at a greater frequency or less than their "true" value.

- The general economy is rife with repossessions, bankruptcies, job losses or general "doom and gloom."

If you notice these trends before they become widespread, you can prepare, but if you bury your head in the sand, your financials will likely take a direct hit, and you may be one of those people who are in danger of neglecting their horses or are forced to sell up (or both). This is known as a fire sale. Not only is this a horse welfare issue, but it reflects poorly on breeders in general.

Aside from serious injury or death, your main fear around horses will be your finances, namely debt and over-extending yourself. So consider what other strings to your bow can provide additional income outside your wages or profession - agistment, training, farriery, massage, rug repairs, leather work, bushcraft, consulting, sponsorship, photography, AirBNB, an online store etc. Just remember all this takes up your time and especially takes you away from your horses.

I know of no stud in Australia that can make a living off breeding horses. All have other incomes which support the stud. However, that doesn't mean they can't pay their way or produce a small profit after several generations.

Building horse-safe facilities on a blank canvas property

From Jenni Phillips, Skyview Stud -

"We're in our 3rd time doing this now (2023). The shortages of labour and materials and price increases following COVID have been significant. We're having to add at least 30% to any contract labour and materials cost from what we did 3 - 5 years ago, and double or almost triple the lead time required for any contractors to be available to get the work done.

"As an example, in labour and materials shortages, our sheds were ordered in (from memory) April 2021, materials were completed January 2022 due to a steel shortage, the shed landed here to clear the shed yard in June 2022. It then rained from August to November 2022 and was too wet to bring the earth-moving equipment in without tearing up the surrounding land. So from waiting again since December 2022, we've just this week (March 2023) had the earth-movers come in to cut the bases, only to be made aware the shed erectors can't be here to put it up for potentially another 12 months. The shed company has now said their build wait list is 18 months + for any new orders plus 6 - 8 months to go through the Council and government planning process (once your design is finalised and you've given the green light and paid the deposit).

"This time it's an old, run down property that was rented for 30+ years, and we're 3 years into a 5 year plan to turn it into something that is comfortable, safe, and has the facilities we would like for breeding a small number of foals (2 or 3 every second year) and working and showing a small number of horses .

"We laboured to clear up all the rubbish ourselves and every job we can without paying someone else. (Ripped out the fencing, loaded and removed many skip bins of household and farm rubbish, removed wire and debris throughout the paddocks, removed the rusted machine bodies, cleaned up and burnt the broken trees and branches everywhere.) It's taken almost every weekend for 3 years in 2 stages. And it's hard physical labour.

"Yes, it can be a labour of love and done less expensively, but if you're breeding *and* riding, there is a balance of this and riding "safe" and having emergency facilities that are safe for the horses and vets / owners. We obviously haven't done it all yet because it's a yearly financial juggle. We're working to a 5- year completion plan, which means there are lots of sacrifices and a HUGE amount of effort in the meantime.

"We contracted the post knocking and had the gates made (which was cheaper than purchasing the gates retail) and did everything else ourselves. The external fencing will also need replacing in the next 5 years which we've just costed (us removing it, knock posts in and put up) - 2 strands barb and ringlock, as we'll run cattle in the laneways at $20 a metre, new internal fencing and water / irrigation (posts, mostly 2 strand Stockguard, small mesh gates, cement water troughs, bases and fittings, 1200m 1 1/4" poly pipe and fittings, trenching) for 30 acres. Total $120k.

"Agricultural shed to take the tractor, tractor implements, horse float, spray and fire-fighting equipment, old ute, and 2 bays for hay. Plus the stable facilities including 3 stables, wash bay, feed shed, tack shed, concrete, rubber in the stables, plumbing and electrical for lighting and a hot water system, and cutting the bases. Plus the 2 rainwater tanks (one for each shed) that in a bushfire zone and with the shed roof catchment areas, are now a council and government building code requirement. It includes us cladding the stable block and welding the 3 stables and panels ourselves. Just getting the front panels and doors made. Total $150k.

"Add another $30k for some form of garage as well for the farm tools or equipment and a cement floor. Cement is around $85 - $90 a square metre now for anyone who is in planning mode. Arena 60m x 30m with a simple fence and adjoining round yard 18m, made of portable panels. Base $35k, sand $15k, + panels, fence $7k, tree removal $7k. Total approx $70k.

"Yes, we are still debating going smaller, 60 x 20m or 40 x 20m, but with young horses the extra 10m width gives greater turning space or the ability to set up some poles down the 10m side or jumps for gymnastic work / head change / a different activity than flat work.

"Pasture rejuvenation - soil test, neutralisation through liming and gypsum and reseeding, add another $50k - $60k over 5 years (has to be staged), to allow reasonable feed on non-acidic soil and hay production with the aim of self-sufficiency.

"So, total cost with overruns, around $400+k.

"Now, this seems like a ridiculous amount of money, but this is where the Business Brain kicks in. If you buy the right block, and add the right facilities, and don't over-capitalise, then you can make money doing this if/when you sell. It is really important to know your market, and how much the property with the increased facilities will be worth, ie capital growth. Plus if you're able to run as a business, the capital expenditure is a tax deduction.

"That means you're not only doing all of this but you likely need 2 other incomes, but if managed correctly, you can claim the costs against the tax you're paying as an employee, or would have to pay as a contractor / self-employed business owner.

"It absolutely means you're double-timing everything for about 5 years. And likely grumpy and tired.

"Which is why it is a labour of love and commitment, if you choose to do it. Not counting the horse work, human kid work, normal family and housework.

"For those thinking "yes but you pay capital gains tax if you use part of the property and deductions as a business expense and then sell the property" - you're right. But if you sell it, capital gains can be rolled over onto the next farming property you buy and the Government kindly allows farmers once they are 55yo+ not to have to pay CGT if they sell their farming property. So there are ways to manage this with your accountant, if you decide to go down this path.

"The point is it is expensive to build proper facilities, even if you cut all of the above in half. It can be done, but it takes effort! And for those agisting at $50 - $100 per week at a place with some sort of decent facilities and wondering why it is $100 a week, and wanting their own single or joint paddock with an arena and some form of possible stable facility if the horse gets injured or needs show prep, and an undercover area to wash off in when it's too hot or wet and for the farrier, we haven't even added in the mortgage costs for the land either.

"Yes, you can absolutely get away with tying a horse to a post to wash it, and we've absolutely done this and are still at the moment, but if you have a horse with a serious injury and need to clean and dress it regularly and confine it for sometimes months, or you have a mare and foal that needs confinement for treatment, that is NOT FUN in the pouring rain and hail and mud, and your vet is likely not to be thrilled either. Nor is it awesome to have farming equipment out in the rain. Ours has been tarped for 3 years now, except when in use.

"Given the expenses to breed, and have some form of "reasonable" facilities, it's not for the faint-hearted, and it's not something you can do for 2 - 3 years and then quit. It's a multiple year, whole family, commitment. And if that's not what you truly want to do, you're much better finding the youngster with the exact breeding, exact sex, exact colour, exact markings that you want, and paying reasonable money for it, appreciating the breeders' time, effort and investment that has gone in to getting it there, and then agisting it somewhere reasonable, understanding why it's not $40 per week to do so. PS: It helps a lot if you have project management experience or love to be organised, and know how to use Excel."

Breed for the love of it, not money.

Financial musts

- Talk to your accountant. Find the best structure to set up your business and what you can claim. You'll need an Australian Business Number (or its equivalent outside Australia.) Do you need to register for GST?

- Your stud should have its own bank account, credit card, etc. Decide who is/are the signatories. Automate. Set up direct debits for recurring bills.

- Public liability, property (including fencing), contents, car / tractor, horse float and horse insurance. Talk to your insurance company. Often you can combine them at a cheaper rate than taking out separate insurances. Ask others who they deal with and shop around. Read the fine print.

- Budgeting. Keep all invoices and receipts.

- Be wise about "good debt" and "bad debt."

- Other strings to your bow.

Why Budget?

You can't budget for the future if you don't know your past.

To track your progress, I suggest you put all your figures onto a spreadsheet if you haven't already. If you are not the "spreadsheet kind of person," you will need to record everything in a folder. The simplest way to do this is to have a plastic sleeve for each spending category and write all your spending as it happens or on a specific day. For the past 30 years or so, it has been Saturday morning for me. Put all your receipts into a plastic sleeve. Then total the columns at the end of the year and give to your accountant.

Only record those items directly related to your horse business. Do the same thing with income, usually sales, but it could be other services, like agistment. I have listed typical expenses which should go on your spreadsheet (or folder). You can also break it down into sub-headings or monthly totals to be added at the end of the year.

Take a long hard look at your past figures. Have you included everything relevant? For example, have you allocated a percentage for petrol, office, stationery, etc., if you have other businesses and income? Can you buy equipment and claim it against the business? Does your income take you into the next tax bracket? Your accountant will guide you in this.

Organising receipts and breakdowns, ready to total at the end of the year.

Look at your final end-of-year figures. Have you spent more than you earned? Where did the money come from to top up your payments? – Was it from savings or transferred from your wages? Did it go on the credit card?

Have you overspent? We have not factored in your labor and support network, as you supply your *time*. Neither have we included any capital expenses (set-up costs, such as the purchase of property, horses, and fencing).

Develop a contingency plan so that if you regularly go over, you know where the money will come from *ahead* of needing it. Then, eventually, after some years, if you are breeding good stock, you will find your horses can support themselves financially. Unfortunately, few people reach this point because they don't survive the first 15 years or so.

Use your figures to tell you where to cut down. Decide *where* you can cut down. For example, you may take your lunch and thermos to a horse show to spend the saved money on entry money. Buy second-hand and not brand-new. Sell off your old gear which you are hardly using. Repair or recycle everything. Old tyres make a great lunging ring. How do you think I know this?

Compare the market annually, for your electricity bills, mortgage and the like. Shop around. Did you know that if you have been with a bank for some time it is worth ringing their redemption department to ask for a reduction in interest rate or "you will likely take your mortgage somewhere else?" They will ask what other rates you have been quoted, so check with other banks first. Go through a Finance Broker for the best deals.

Expenses Year	Year Totals $
Advertising and Marketing	
Agistment	
Branding	
Breaking / Training	
Clothing	
Dentist	
Entry fees	
Farrier	
Financial (accountant fees etc.)	
Float expenses	
Fodder	
Gear/equipment	
Genetic testing	
Insurance	
Legal	
Memberships	
Office	
Petrol and car (%)	
Property Expenses (divide this up)	
Registrations	
Stud Fees /semen purchase	
Subscriptions	
Transport	
Veterinary	
Other	
Total $	

Add the totals from your tax invoices and place them here at the end of the year.

Aim to avoid "Bad Debt". Only take on "Good Debt". Do you know the difference? Good debt is a loan towards something that will increase in value while you are paying it off, such as a mortgage on your property. Most well-located properties will double in value over 10-15 years, resulting in you having gained equity over that time. You can use that equity to borrow for improvements on the property, or to buy a better property.

Bad Debt is where the security, such as a car, is worth way less when you sell, than the original loan. So, I wouldn't buy a car on credit, especially from a car finance company, as it will charge more than a bank. On the other hand, if you buy your horse float and take good care of it over the next ten years or so, you will most likely sell it for near what you paid (and possibly more). A loan taken out to purchase it then, would be "good debt."

It is wise to set up a savings plan as a financial "buffer." Use a separate dedicated banking account and put regular amounts of money in by direct transfer before you even see it and spend it on something else! As they say, pay yourself first. It may be as little as $50 a week from your wages, but anything is better than nothing. Next, consider investing the money into an interest-bearing savings account, which locks you in for, say, 3-6 months, or longer. That way, you don't use the money for trivial items. If you can't find at least this amount, I recommend you hold off breeding horses until much later. Discipline yourself so you never use it except in a dire emergency.

One way to look at all your figures is to decide how many horses you need to sell to pay your annual feed bill. For me, this is one horse. Do this for all your expenses. It turns the budgeting process into practical understanding and motivates you to continue budgeting. It tells you how many horses you must sell each year to stay afloat. There is no short cut to this, and no pot of gold!

Let the figures do the talking.

Arakoola Australian Stock Horse Stud youngsters. Photo Janet Hodges.

Chapter 7

Your property

White rails, a tree-lined driveway, and a large barn do not make a horse stud. Neither does one with barbed wire yards strewn together, rubbish everywhere, and horses standing in manure. The first is often associated with glossy advertising and the other with Facebook posts. Both can attract the wrong kind of buyer. There is a saying that the buyer who comes in the driveway with the most expensive car and float wants the most unsuitable or cheapest horse and is the biggest time-waster. In contrast, the one with the old rundown car, stepping out with jeans and a well-worn Akubra, is prepared to spend the money on a good horse and knows exactly what he wants. Although this is a joke, there is an element of truth.

What makes your property a reputable stud is what visitors see of your horses, their quality, care, and handling. So first impressions count. It must be neat, tidy, and safe for you, your horses, and your visitors (and staff if you have them). But it doesn't need to be a showpiece. Many of the older studs are well-worn, but the minute you set foot in them, you know you are part of a timeless tradition of great horsemanship and excellent horses bred with a purpose.

You don't need to own property to breed a few horses, but it helps. I have bred horses for over 50 years and only spent 15 years on land I owned. Agistment does have quite a few advantages, such as being mortgage free and not being responsible for set up, basic costs, like rates, power, and upkeep. Usually, there will be someone living on-site. It may seem expensive, but not so much if you factor in the costs of running a property in the long run. If you find the right place, it can be ideal for the small breeder. But it doesn't give you security.

Buying a place if you've never been a property owner or a farmer can seem daunting. You need to consider the size and location carefully, and sometimes you are left with minimal options based on work or other requirements. Remote locations are difficult to attract clients.

The ideal is to get the largest acreage you can afford near a known equestrian community. Do you want a lifestyle location and fewer horses, a larger property with less focus on lifestyle, or both? It must be emphasised that the more naturally you can run your horses, with full-time grazing most of the year, the better your horses will be. This will reduce your expenses, particularly for fodder and vet, usually your highest costs. If you can run them with sheep and cattle for rotational purposes, all the better. These will eat down the longer rank grasses, and reduce worm infestation. .

Your agent or agronomist can advise the stocking rates for certain locations. Calculating your stocking rate (or DSE / hectare) is also possible with some online calculators. Generally, about 10 sheep equals one horse. Remember many factors, such as seasonal conditions, pasture health, rainfall, and supplementary feeding, will influence this.

Check your water supply and its salt content. Bats, processionary caterpillars, parasites, and mosquitoes are other factors to consider for disease control, so know the specifics for your area. If you don't learn about them now, you may find out later at a cost.

You don't need a perfectly flat property. This may benefit staff, but hilly or undulating is more beneficial for your horses. For health and conditioning, the athletic horse is better off being raised in good hilly country, with well-timbered breaks, than in small flat paddocks or, worse, in stables.

A typical Aussie scene, a contented mare with her foal on undulating country with plenty of trees. Chalani Anna and foal. Photo Peter Gower.

If you are buying a place from scratch, it must be designed around both horses *and* family/staff. I see many expensive designs which are wonderful for the horses but very labour-intensive for staff. The most labour-intensive routine activity on any stud is the feeding ritual, once or twice daily. If it takes a full hour to do so, it is hard to make it any quicker if you're in a hurry or something urgent comes up. Because it is at the end of the day when you're exhausted, want to finish, go inside, rest, you can't because it's feeding time!

It is often best to live on a property before designing it from scratch, as you will likely change your mind or wish you had done something different once you are working on it. Start with a map, walk around and mark out where you might place everything. Then go back and redo it with specific measurements. You may need to do this multiple times before you are happy with it. Ask for at least two quotes if you are contracting.

An ideal setup is the wagon-wheel system, whereby you use your house or your main shed as the centre of the wheel, and your paddocks radiate around the centre. The smaller yards and facilities are near the centre, with progressively larger paddocks further out. The less walking you need to do to catch horses, the better. Then design laneways for the further paddocks, which are easy to walk or take a vehicle. You can also plant trees in the laneways, knowing they will be protected from the horses. Ideally, there should be double fencing between paddocks to keep horses from each other, as this is the most common cause of injury. Watering points should be as near as possible to the wheel centre, so you don't have to walk too far to check, preferably with the float outside the fence.

The most expensive fencing is not necessarily the safest. Wooden fences require much maintenance but may still stand in 100 years. Plastic fences are certainly not long-lasting and may be a waste of time. If you are going to use artificial products, research them carefully. Some are exceptionally good. Never use high-tensile steel wire. Ideally, no wire, but whatever you use needs to break relatively easily when a horse has some part of its anatomy stuck in it (e.g., gripples or end connectors that will "let go" under pressure). Droppers are highly dangerous and must be capped to prevent horses from staking themselves. Double wire caps are better. Best not to use them at all, as the caps come off easily or deteriorate.

Find a place where your horses like to stand and put the shelter there. Big open or one-sided shelters are best for a group. Otherwise, put the shelter near a gateway so that it is easy for staff to place food in it. Better still to have a dual entry from two paddocks because the horses will want to stand where there is company.

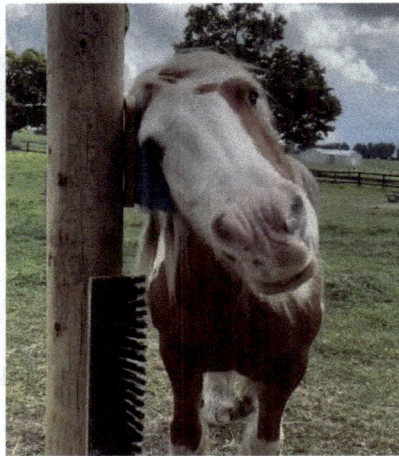

Homemade scratching post

Paddocks with few trees need shelter and scratching posts. If you don't supply a scratching post, they will rub on what little they have available, usually your trees or gate!

Babies will go over, under, and try to go through stranded fences. Areas where you graze foals and most yards will need "foal proofing." If the existing fencing is not up to scratch, electrify it until you can do better. I believe that all horses are better off behind electric fencing if you can manage this.

If you're considering a stallion, house him close enough to be ridden and near where you will serve with him. It is best if he can have a view over the whole property or at least some of the daily activities. Allow room so he can have a gelding or mare in with him for company at certain times of the year. If this is not possible, have a companion nearby. Even a "mini" in a laneway next to him is better than nothing. You can set this paddock up with higher fencing. Many stallions will run the fences, so the happier they are, the less likely this will happen.

Gates should swing inwards to restrict others from running through when you are bringing one out. Latches and bolts should be placed with care and all danger points located and minimised. You may need double gates for implement access. Never have a gap between the gate and fence post large enough for a horse to trap its neck. If you are living on the property yourself, it will likely be a work in progress to set it up exactly as you would like, so factor in priorities and costs of what is essential and what can wait.

Consider whether you will grow your own hay, its storage away from other buildings in case of fire, access, and how you will feed it out.

Select pasture types known to be suitable for horses, and especially be careful of those which contain a lot of clover, as some clovers are notorious for affecting fertility in mares. Do you have enough paddocks to allow for spelling and rotation? Plan to rest paddocks and slash longer grasses. Encourage dung beetles by avoiding chemicals. Keep these in mind when purchasing your property. Know your local area.

Another labour-intensive activity is manure collection. This is imperative in all stable/yard areas and should be done at least weekly for larger yards. Where will you put the manure? Big paddocks may need a ride-on vacuum system; larger ones should be harrowed and spelled. Soil pH and weed control are adversely affected by manure left on the ground, and some weeds are toxic to horses. Soils can quickly become very acidic and if so, will require annual liming. The biggest problem with intensive horse holdings is pasture degradation by hooves and general mismanagement. Run-off will affect creeks and waterways. Check your council regulations for stocking restrictions and with your local agronomist for advice.

An excellent investment is portable yards. They are perfect for renters and property owners alike, and you can move them for any purpose. They can be used as temporary fences, gates, lanes, or lunging rings. We like to use them as foal-holding yards in our broodmare paddock. When we rotate the broodmares, we shift the yard. They can help with closing up a horse overnight or weaning foals. They have so many uses you will wonder how you did without them. You can buy them before you even set up your property, which helps you determine where you will put your permanent facilities.

As you can see, most of this work requires a good tractor, so make sure you factor this in as an essential purchase, when purchasing your property.

The type and number of horses will dictate what handling facilities you require. For example, you will need more small yards for ponies. Better to use a horse "track" system, where a track is built around a main pasture paddock with slow feeders on the track. The ponies walk around the track most of the day for exercise and stimulation and are only let into the paddock at night or for part of the day.

Vet facilities may be power, lights, and water, with a crush, right through to full kitchen and lab facilities for artificial breeding, next to serving facilities. An office can be as simple as a desk in the corner of a room with a computer, printer, trays, and a filing cabinet. A dedicated room is far better, with enough space to expand as you grow. You can incorporate this with a reception area, chairs, bookshelves, storage, memorabilia, trophies, and coffee/tea-making facilities.

Depending on how you receive visitors, this can be at the stables or the house. You can even put out a visitors' book. Visitors need access to clean toilet facilities. It is best to make your office as comfortable as possible (air-con/ heater/ window view etc.) and secure, so that you can attend to all your office work promptly, away from distractions, but still be somewhere central to the activities of the stud. If you don't have a dedicated reception area, ensure you have a space in the kitchen or other room in the house or stable block suitable for the purpose.

Typical Aussie scene providing a wonderful upbringing for young foals. Buckajopark Lyndal HSH and foal. Owned / photo by Amy Curran.

One of the best things you can do when you move onto a new property is meet your neighbours. Get to know who the local tradespeople are and the "go-to" persons for advice. Neighbours will usually be only too helpful in this regard. On the other hand, don't upset them unnecessarily with your rubbish or roaming dogs. Friendly neighbours who know your patterns can sometimes be your extra eyes on the place when a horse jumps the fence, or a mare has just foaled unexpectedly. We have found them invaluable.

There is an outstanding website for information on different layouts and systems and healthy management of horses and property. There is no better place to start than this website: Equiculture https://www.equiculture.net/

Your property need not be lavish but it must be neat and tidy. It must project care.

Glorious sunset over a well cared for small property. Both are Skyview Stud. Photos courtesy Jenni Phillips.

Chapter 8

Your foundation stock

Choosing the right foundation stock can be a lengthy journey. Be wary of people who see you as a collector of horses. They will increase the price as they see you coming or sell you something they want to be rid of. This is well known to happen at auctions. I would not recommend auctions for the novice breeder, as it will be too easy to be talked into buying something high-priced or ultimately unsuitable.

Study your breed and its rules. First, you must understand your Breed Standard, then select animals as close to this as possible. Next, talk to your mentor, especially about which qualities are the most important and the current "fashion" that may have departed from it. The standard doesn't change "with the times," and there is a reason for this.

Read your breed magazines and visit studs that interest you. Read the stories of the greatest horses of your breed. These give you a head start to the lines you like. Who are the breeders, trainers, and performers in your industry? Who are the most reputable in the industry? These will not necessarily be the ones making the loudest or biggest noise. Speak to the "old timers." They will have a lot of knowledge about older bloodlines and their qualities.

Understand pedigrees and bloodlines, and familiarize yourself with popular names and, importantly, other less well-known names.

"Well bred" refers to a horse having a good pedigree filled with performed animals, as well as "pedigree depth" where the animals in the further generations are also known to have performed well in the field you are seeking. These need to include consistent family successes rather than the once in 5 generations successful freak. "Well bred" means its forebears are known to be sound, sane, well-conformed and useful. At least some, preferably a lot of them, will be highly successful in the field you require.

Decide how many broodmares you will need per your property's capacity. *A five-mare band is full-time work for one person.* Five broodmares will likely mean five foals, five yearlings, and five 2-year-olds awaiting breaking in. In addition, there will perhaps be a stallion, one or two riding horses, and a retired horse. That alone is 23 horses. Then add any further extras you might have. Can your property and your workload support these numbers?

Finding breeding stock in the prime of life will significantly raise the price of what you can buy, especially as you want something well performed. It may be better to purchase stock past its prime but proven to produce top-class progeny. Your mentor, research, and networking should provide a shortlist of where to buy. Explain what you are looking for and your price range. Be specific. Ask to be referred to someone else if they don't have what you need or put in an expression of interest in something that might become available in the future.

Mares should show quality and femininity. Skyview stud's foundation mare Goldmine Champagne Blush HSH. Photo by Karen Sheridan.

Price is not a good indicator of value. Don't believe the lie that the more you pay for a horse, the more capable that horse is, when that is only sometimes true. People are just unwilling to step out of traditional concepts of where a "good" horse can come from. Thoroughbreds are a great example. They cost so much at the beginning of their lives because you are buying the story and the potential. Unless they prove that potential through huge earnings, they cost little to nothing when they are done racing because the story has changed. They are the same horse.

On the other hand, don't expect to find good breeding stock which appear to be cheap. There is always a reason why they are cheap. The horse breeding world is filled with these, and their progeny generally amount to nothing.

Try to diversify the bloodlines with your initial purchases. You cannot possibly know which foundation stock will be your best down the road, and you may want to keep fillies from all of them initially, at least those which measure up. I give a broodmare three chances to breed a foal up to standard (or better) before deciding if she will continue in the stud. Over time you will find that one or two mares are the better producers, and your stud will consist mainly of their progeny.

Leasing a good mare is another way to acquire good breeding stock. Mares are readily available for free lease to the right home, but if the owner values them, there will be conditions. Make sure you get a contract. The lessee is responsible for all care, vet treatment, and subsequent bills, as if they were the owner. Have the lease recorded with your breed association; then, you can record the foal in your prefix. I have leased multiple mares over the years, never with a problem, but over time, only two became what I would call foundation mares whose descendants are still with us. Three to five generations from our original mares, our stud traces back to just three, two of which we leased. Others we have bought have not been in the stud long enough to see if they will have generational influence.

Buying an aged broodmare can be a cheaper alternative, as long as the mare has recently had a foal or is in foal and looks in good condition. Look at her breeding history. Choosing an aged maiden mare is risky as you may never get her in foal, and it may be an expensive exercise. Problem mares are not worth considering if you are starting. You will quickly spend dollars trying to find the problem and may still not get a foal. You might also be able to pick up a close relative to a well-performed individual. Healthy mares can frequently manage 10 foals in their lifetime, and some many more.

Aged, well-cared-for mares often breed into their mid-twenties. They will usually bloom when pregnant because of an increase in hormones. Horses have a long reproductive life and a very short "old age," unlike us. They will frequently die of natural causes only two to four years after ceasing to be reproductive. Occasionally, mares have produced when 29 and even 30. I knew a maiden mare that foaled at the age of 28. They must have an excellent feeding regime, especially with a foal at foot.

Have all mares vet checked for breeding soundness if they are not currently in foal, and check the carrier status for genetic disorders (see chapter 18). Your breed association will have this on record. Otherwise, pay for testing in accordance with your breed.

Pick horses with suitable temperaments, especially if you are a novice breeder. There is so much with which to contend without adding the wrong temperament into the mix.

Work out your non-negotiables and must-haves. What are your standards? Take your time to get the right stock. Aim for excellence, but not perfection, as no horse is perfect. Instead, aim for the best you can afford. Know what faults you can live with to find the best available in its price range. Trying to breed something "exceptional" before establishing a baseline is a mistake. This is where using foundation stock from well-developed older studs and bloodlines is invaluable.

Now set up your strategy based on the results above. Use the first person for greater clarity. For example:

"I want to breed from two mares per year and keep fillies from each as they become available, only if they are better than their dams. I will sell the rest, including all geldings. I will send fillies to the breaker, then train and perform them myself. I can work with a maximum of two at any one time. The best ones will retire for breeding. I want to build this up over 10 years to five breeding mares producing five foals yearly, then sell all bar one pre-breaking in. For every animal I keep, I will sell another."

Time is on your side. There is no rush.

Finding quality breeding stock can seem slow, but the time spent now will significantly save you years in the future.

Chalani Sheoak, a combination of two Chalani foundation mares. Photo by Peter Gower.

Chapter 9

Conformation and breed type

"Choosing individual stock without any idea of what you are looking for is like running through a dynamite factory with a burning match. You may live, but you're still an idiot." – Joel Greenblatt

Having a practiced eye

Conformation means having the correct *form* for the horse to *function* properly. This varies slightly from breed to breed, but the basics are always the same. At Expo 1980, I witnessed two of the greatest Australian mares of their time, Clover Bonita (QH) and Desert Queen (Arabian), standing side by side. What struck me was how similar they were, not their differences. No feature was extreme. All was in balance; they were feminine and quality oozed from them. They were wonderful representatives of their breed.

I can't emphasise enough the importance of sound con*form*ation and knowing your breed standard. What are the distinctive qualities of your breed? What are the common faults which owners of other breeds notice? (Don't speak to the converted here).

Function follows form. Soundness and form are interdependent, and breed types develop through function and purpose. Whatever the discipline, there are basic, tried, and true foundations of conformation, each with relatively small room for error without upsetting the mechanism. There may be trade-offs and compensation mechanisms, but rarely will there be extremes performing well in such disciplines. This is why it is important to know if you are following a "trend" that is unnatural and pinned by judges or staying with the proven rules of conformation and your Breed Standard.

How do you make all but the most superficial selection decisions if you don't know what is important and what isn't? This is not about taste, personal opinion, what is in the eye of the beholder, or even what the judge thinks; it is about standards. These standards are found in any good book on conformation and are readily available for anyone to learn.

Try to breed to the Standard of Excellence of your breed. Chalani Galaxy HSH as a yearling.

What makes a horse a leg mover rather than a back mover? What shape gives a horse the most flexibility? Why does a horse with a table-top croup tend to drop out of its canter or dis-unite? Where is the LS (lumbosacral) joint? What is a good saddle back? How should a neck be shaped? Where is the horse's centre of gravity?

I started by reading books on conformation, then drawing over magazine illustrations. I drew lines over the illustration for every fault I found to indicate where the horse "should" be. For example, I drew a line through the shoulder to illustrate the angle of its shoulder. I drew lines through upside-down necks to where the line "should go."

Look at your own and friends' horses. Go to sales, talk to judges, or go to the races. Speak to old-timers. Go to judging clinics. (This doesn't mean you will need to become a judge). Look at a lot of horses and get feedback. Actively develop a practiced eye.

Developing a practiced eye is not hard. It just takes effort. Doing it with a living animal is harder, but it doesn't take long if you approach it with consistency and excellence in mind. Without knowing your conformation, you will never breed top horses. You will be breeding on pedigree alone, and not all horses live up to their pedigree.

Extremes of fashion go out of style for the next fashion. Don't chase trends for the sake of higher prices. Always seek a balanced horse. Then if it has a slight leg fault, it usually doesn't impact its usefulness or the rider's comfort. On the other hand, you don't want to go so far in one direction that you sacrifice something else. An example might be going for speed in the racehorse without consideration for sound feet.

At events, you will see ones that out-move or out-show yours, so you pay attention to what is winning and then let it influence your breeding choices the following season. Judges may be blamed for their "poor" choices. Surely these winning breeders know what they are doing, so you follow their lead by picking younger flashier stallions.

Before long, this creates an echo chamber where people are breeding for a flashy trot and "pretty." We start to lose a very rideable, versatile, and seriously sound line of horses because they take a long time to mature and thus are not sellable as youngsters, and the "flashy" horse seems the most marketable. Over time, the breed develops two different types: the "show" horse and the "not so show" horse.

There can be major incentives and prize money for breeders specializing in futurities. Unfortunately, many owners disagree with these, and I feel that such money should be offered for adult horses, not babies, especially not two-year-olds. Trainers and owners follow the money, so it should come as no surprise that older horses get sold off or returned to the breeding band without showing their full potential, thus increasing the volume of horses on the market with an uncertain future.

Breeders may say they breed for the "average rider" but that is about its temperament, sound mind, and selecting the breed that will do it for the average rider. It does NOT mean breeding average, or poor types of horses. The average rider, even more than the pro, needs a sound, well conformed horse that has been well trained; a friend that he can have a partnership with for the next 20 years or so. We must always breed good natured versatile horses, without exaggerated movement or faulty conformation if we are breeding for the "average" rider. Even the "average" rider, likes a horse which can be admired. The average rider knows what he wants and will usually pay well for a good type of pleasure horse.

I hear people say they breed working horses, not show horses. But the working horse must still be good to fulfill its role. Maybe it doesn't need a pretty head as you don't ride the head, but it does need to be fluid, athletic, hardy and workmanlike. It is less likely to be stressed and fatigued if it is well balanced. A lot of performance horses are picked up at auctions. If you want to sell at auction, you must consider that the best-priced horses have pedigree AND conformation. A good-looking, well-conformed horse will always sell.

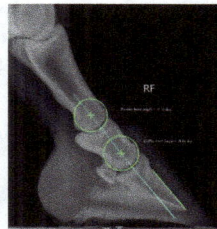

Ideal Alignment

RF

Reprinted with permission from Progressive Equine.

The foundation

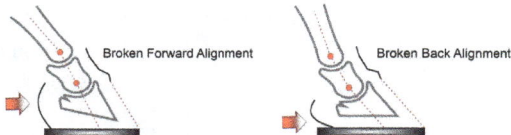

Broken Forward Alignment

Broken Back Alignment

Which is the club foot? Which one has the crushed heel?
Once you see, you can't unsee it

It is important to study conformation in theory and the living horse. You cannot learn enough about conformation (I include movement). Farriers and vets will talk about legs and soundness, which is the most important characteristic. Good hard hooves with deep open heels, should be in alignment with the slope of the pastern and shoulder. Upright pasterns (greater than 45 degrees and 50 degrees in the hinds) will nearly always indicate an upright shoulder and pelvis. Noticeable faults are turned out or turned in feet, but the less obvious ones, such as offset cannons, calf knees and "tied-in below the knees" are much more severe in their effect on soundness, yet easily overlooked by inexperienced horsemen. They are also faults hard to breed out.

As a breeder I value structural integrity and its role in soundness. It's unfair to the horse, the new owner and the breed if you produce a foal that will develop unsoundnesses later in life because of some obvious or genetically possible structural defect. Worse if breeding from one that has broken down due to such defect!

Some judges only assess a horse for its faults. Unfortunately, the fault picker will be so critical that he often misses the good qualities. This results in an average horse. Sometimes the only good thing about it is its balance. It is very plain when you see it next to a quality horse. Do you want to breed average?

The only time it is acceptable to use an average mare is when she is a proven, personal favourite; even then, she may not be breeding quality. All your other mares should be of a high standard. Always aim for excellence.

My horses must have a good head and a soft eye. This indicates intelligence and character. There are so many horses bred that it is easy to have this as a requirement and not add to the ordinary heads in the population. It is the first thing I see when I feed in the evening. And it makes it easier to sell horses because they have more appeal. In mares, I like to instantly see femininity.

The horse is the sum of its parts. So first, look at balance, examine each part separately, and then reconsider the parts as a whole. How does that affect the horse's ability to function and stay sound? Does a line drawn around his body form a square or a rectangle (portrait or landscape)? The horse should be long on the underside, and short across the back and loins.

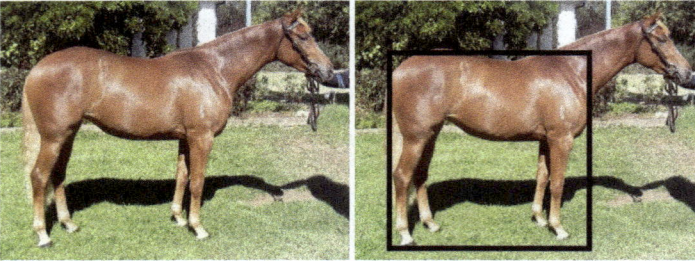

Balance: Everything is harmonious and in proportion. This is a very well proportioned 2yr old, just a little high behind as youngsters often are. Note the elbow room. Imagine a square round the body.

Some qualities come down to personal preference. Mares tend to have longer ears than the same genetics in stallions. So if the stallion has long ears, his daughters will likely have longer. When both parents have long ears, the progeny may end up with "hat or floppy" ears. I like them wide between the eyes, a flat forehead, and no bumps along the nose or forehead. I tend to like a horse with whorls, as so many times these horses have more brains and quality.

I look for a clean gullet. It might be thicker in ponies and the baroque breeds, but it should be there. There should also be a shadow line down the length of the windpipe and the underside of the neck should be curved, not heavy or upside down like a banana. I'm not fond of straight necks because they are nearly always weak at the poll and hard to collect, but this does not impede the racehorse. He just needs the shortest distance for the air to enter his nostrils and reach his lungs. The horses next have an excellent gullet, neck, and windpipe line. Make sure that you look at them naturally, not posed.

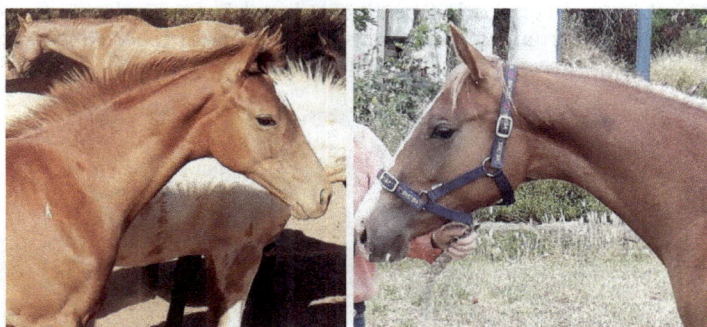

The gullet should be lean and open, with a defined line down the wind-pipe/jugular and a concave curve under the neck.

A neck set is dependent on the slope and length of the shoulder. A low neck set leads to a shorter, more choppy stride; a high neck set leads to higher knee action or a gaiting stride, often with a hollow back. Horses with raised or exaggerated necks are the gaited breeds because this enables the higher steps or the gaiting. (Pacing Standardbreds wear an overhead check to raise their neck). If you don't want a horse to "gait," go for balanced movement and a neck that attaches to the head, where a line drawn from the withers cuts through the head midway below the eyes.

Where do you put the saddle? What would it be like to ride either?
Which parts of the body are most likely to experience wear and tear?

Horses with straight necks or downhill conformation are usually heavy on the forehand, shorter striding, and harder to get softness without an experienced rider. They often are more uncomfortable and less supple. This tends to lead to riders heavy on their mouths, jamming them up, or simply riding them on the forehand all their life, lessening longevity and soundness. A horse with an apparently long back, is usually caused by an upright shoulder with a weak loin. It tends to have a strong trot, but a rough canter. An overall rectangular look may be due to shortish legs.

Hindquarters should have length, from side on, and good hams and gaskins. The gaskins, viewed from behind, should show strength both inside and out. The first photo on the next page, shows "post" hocks. Post legs are where the hind legs are too straight, giving the rider a feeling of sitting on a pogo-stick. They are highly prone to luxation of the patella, most commonly found in Thoroughbreds, Quarter Horses and Minis. The second photo shows a tent rump (^ shape) and asymmetry, which indicates injury or soreness. This makes it difficult for horses to reach their hocks under their body easily. Damage is a feature of many jumpers, and is the cause of the jumper's "bump". Photo 3 is correct, where the muscles either side of the stifles are slightly wider than at the hips. This is the power house, the amount of muscle dependent on breed. Last, a good type of hindquarter on a sprint horse. Tent-rumped horses will suffer strain with collected work, while the horse with the right shape, will strengthen further with work.

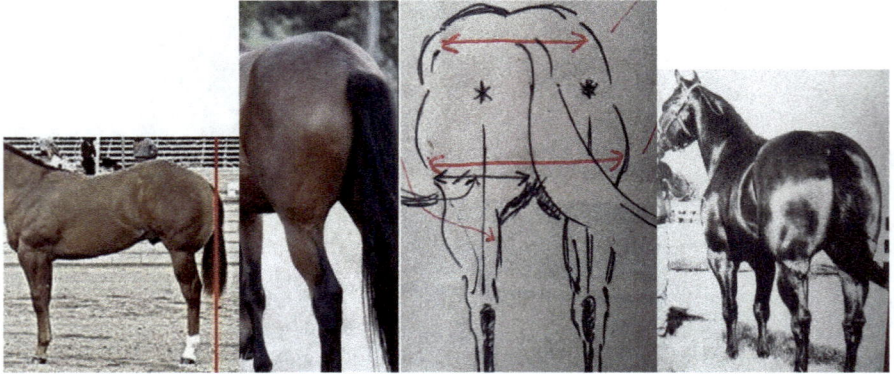

Different shaped hindends. left to right: goose rump, tent rump, correct, and correct.

Note that sickle hocks (hocks too bent) are not to be forgiven simply because they allow the horse to stand under its body. The extra strain can lead to unsoundness in the joints. Good flat, not fleshy hocks, are essential for any horse in regular work.

One of the unmentioned "secrets" of the horse world is the amount of illegal substances used to keep a horse sound long enough to perform. It is a disgrace. Breeders and trainers need to do far better, together with rules and policing changed within the disciplines concerned. We could blame judges, trainers, the vets or Big Pharma, for using the procedures and medications they have developed, in particular, the modern trend of injecting joints to keep horses competing. Draw the line between ethical intervention and keeping a horse going for one more competition! Unfortunately it's often about money and the present, with little regard to the future. There's nothing wrong with managing a retired horse using your vet's expertise to do so, as long as you are doing it for the right reasons and with the horse's welfare paramount.

So make a list of your non-negotiables in your notebook.

Here are some of mine: Crypt-orchid, parrot mouth, turned feet, poor hooves, offset cannons, tied at the knee, calf-knees, cow hocks, upside down neck, downhill build (wither lower than hindquarters), tent rump and hindquarters lacking length or depth.

These are my must-haves: Quality, balance, good legs and hooves, flat straight hocks, square stance, good head, gullet and neck, sloping shoulder and distinct wither, deep girth with good elbow room, well shaped back to carry a saddle without sliding forward, back or sideways, strong over the loins, and the right movement to make a good saddle horse for comfort and durability.

For me, quality is in the fine skin, the lack of "feathering" on the legs, the refined head with a large, kind eye, broad forehead, silky mane, and defined muscling, strength with elegance. It is in the bearing of the horse when it moves. It is presence! When it all comes together in a balanced and smooth whole, there is one word for it "quality!" Quality may be hard to define, but you know it is there as soon as you see it. The dictionary says it is a distinctive attribute, degree of excellence, or higher class. Quality is not only an important characteristic in your horses, but also in developing customer satisfaction and loyalty. Quality resonates with buyers who are prepared to spend the most money.

Conformation Comparison: – By Kimberley McCallum

"This is what a true fugly is. It is 100% better than when it arrived, but it was never correct, so it will always be fugly, no matter what. Other than the things that are wrong in the profile pic (which many people may not think is that bad, depending on experience) she is also the only horse I have ever come across with offset cannons on both front and back legs. Her hooves were so small she couldn't hold a pony sized sideline, and her neck was physically incapable of offering any vertical submission. She is 2yo here. She is also overshot in the jaw.

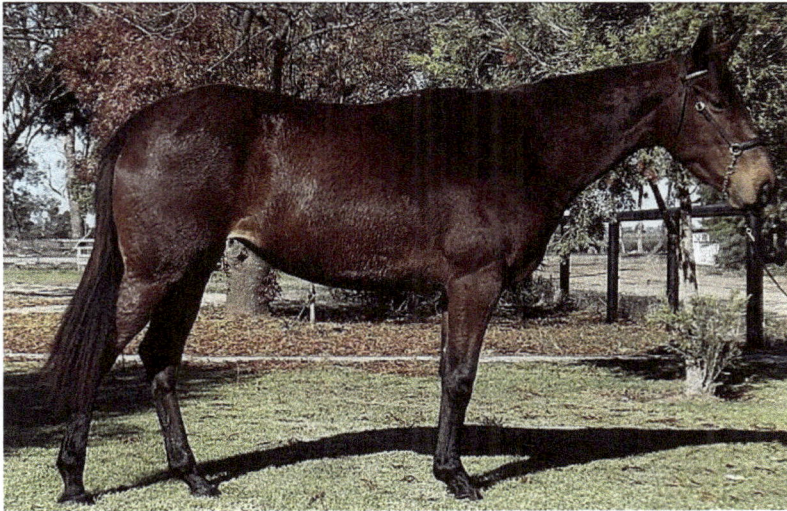

"We have all seen poor, malnourished, hairy young horses turn into lovely useful creatures, but these were always structurally correct. You just have to give a little leeway for the chest to thicken, hindquarter to fill out, etc. Please note the difference here. Although skinny, wormy, hairy, dehydrated and stressed, this horse is reasonably correct. There is 12 months difference in the photos, roughly 2yo in the first."

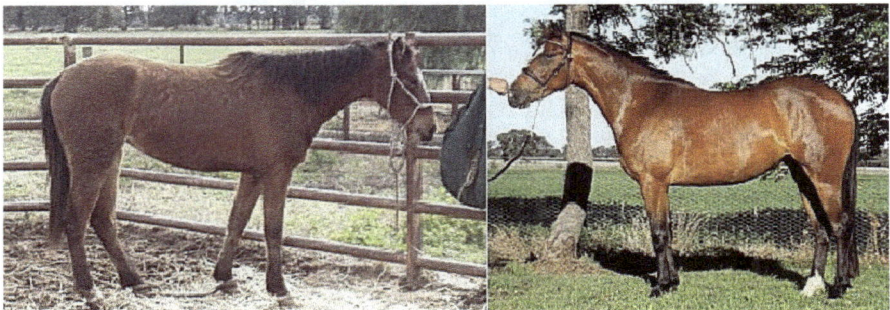

Horses were bred tough in the Waler period. Australian horses were considered the best remounts in the world. Third Light Horse regiment, WWI. Sadly only one was allowed to return to Australia due to our strict quarantine laws.

Disunited canter - this is not a good look. The hocks are not under the body carrying the weight. There is limited separation of the hind legs, hence the front is coming down too soon. The horse will likely do a "bottle turn." This type of canter is "proppy" to ride. Compare the same phase of stride with the Black Caviar photo in the next chapter.

Chapter 10

Movement

An extension of conformation

"There is hardly anything in the world that someone cannot make a little worse and sell a little cheaper and the people who consider price alone are this man's prey". – John Ruskin

Movement is *very* important. Some consider it more important. The reasoning is that if the movement is correct, the conformation is proven. However, it is easy to find horses which are faulty in the legs or have nothing to recommend them except their movement, so you need both.

Straightness is desirable, but a slight winging out may be acceptable. Correct front legs tend to be more impactful for basic soundness, but the hind legs for more advanced training. Turned-out front legs are worse than pigeon toes because the horse may strike the opposite leg while moving.

Imagine a central line running through the horse as it moves, front on. The faster the pace, the more the foot will fall on that central line. That is normal. The shoulder is not joined to the torso of the horse by a bone like our collarbone. The muscles form a "sling" around it instead, so we want to see plenty of elbow room and a shoulder that is well-angulated and flattish. The footfalls should be light, not heavy on the ground.

The spine has no lateral bend except in the neck and tail - maximum bend for turning is achieved by crossing its legs.

When viewed from above, you will notice that a horse can barely bend its spine between the withers and the tail. To turn, the neck bends in the direction he is going, and he uses this "sling" apparatus to allow the legs to cross in front of one another. Only then can he do a graceful turn. A bottle turn is very wrong, with the horse too likely to fall.

Looking at the horse from the side, one counts its footfalls and the evenness with which they fall.....1, 2, 3, 4. Repeat, repeat, like a metronome. If there is any dysfunction, the horse might be lame. If I hear 1, 2 ... 3, 4, I know the horse is ambling, which for the athletic horse, is bad. It usually indicates stiffness in the lumbar, pelvic, or neck area. Foals sometimes do this as their legs are longer in proportion to their body, so they might grow out of it.

Worse is a trot where the diagonal footfalls are not together, shifting from two beats. This is usually caused by hyper-mobility of the limbs, high "extended" knee action, and a difference between the stride length of the forelimbs and the hind limbs. Hind leg reach and foreleg reach should be of identical length. The selection of hyper-mobile movements by judges and breeders only worsens this problem to the point where some breeds are at risk of being completely dominated by such incorrect movements.

One goes through the same process with the canter: 1, 2, 3, - 1, 2, 3, Repeat .... Faults are the disunited canter and four-beat canter (like a standardbred canter) and amble canter where the front hoof comes down too early. The last leg to hit the ground should be the leading front leg, which should come with great stretch through the shoulder. If the hind leg is not strong enough to hold the canter (often due to sickle or cow hocks, or being too wide behind), the front leg will come down too soon. When looking at a horse cantering naturally out in the paddock, I watch whether it is smooth and balanced, especially on uneven or wet ground, and whether it changes stride and/or disunites as it changes direction. It only took me a couple of years practice, watching and listening, to recognise footfalls as a horse approached and know what its gaits and anomalies were, merely from the sound of its steps. You can do the same.

Separation of the back legs, rounding of loins, strong hams and lowering of the hindquarters. Black Caviar at her best.

Look for a big overtrack and head bobbing at a free walk (or rolling walk). In the stock horse world, it is called the "mustering walk" for riding out over long distances with energy. Overtrack is where the hind legs step over the landing print of the front legs. This indicates the length of stride. It is one of the best qualities a horse can have naturally, as it is very hard to improve the walk with training. It is also one of the best indicators for potential speed and the quality of the gallop. If you want a quick take off, such as in a roping horse, exceptionally strong hindquarters, with hind legs standing well under the body, are essential.

The horse is one of the few species, along with tapirs and hippos, that are hetero-lateral. That means they "trot" rather than "pace". It also means they are built to canter "united", not disunited, like cattle. This leads to a a certain type of conformation - the square and high hips of cattle for instance, as opposed to the horse. There are variations of course in all horses, either by breed - square-gaiters compared with the pacers, or by type, examples being the Tennessee Walking Horse, the Paso Fino, the Icelandic, or the Marwari horse. These differences are inherited.

Gaiting breeds - Marwari horse Gajraj, owned by Sh Gurtaj Sandhu,
Mann Horse Photography. Tennessee Walking Horse, public domain.

In some breeds, such as the Australian Stock Horse, it is a serious fault to pace or amble. An ASH must have a true four beat walk. Poor gait can be a sign of soreness or tension, if the horse travels with high head and an accompanying hollow back or hollow neck, or both.

When judging, I often see horses whose heads are held in a way they cannot properly head bob, and the gait at the walk is ruined. Just as bad is the horse that is "bridle-lame" at the trot. The crooked rider "creates" the disunited horse. Some horses are prone to dis-unite, or amble canter which are two of the worst faults a riding horse can possess.

With movement, one looks at the spring, power (the impulsion), rhythm, and tempo. Such horses will be fluid and graceful and very comfortable for their rider. Correct conformation is necessary if you want your horses to last and if you want them to be comfortable for a long day's work. We all know horses which are exceptions to this, but they shouldn't be used in a breeding programme.

Over time, the best-conformed horses will rise to the top, so long as the breeder and the breed value performance. The highest sales are for those breeders able to produce both the good-looking performance horse and the top-performing show horse. It is not either/or. I thoroughly recommend any of the books by Deb Bennett, or this website to explore conformation further, especially for discipline-specific breeding:

JW EQUINE   https://www.jwequine.com/functional-conformation/

A practiced eye is hard to develop unless you see a lot of horses. So, after reading this short explanation of conformation and movement, I hope you are inspired to learn as much as you can about it. After all, if you want to breed better horses, your future depends on it.

"The scarey part is the unsound stud or mare is bred, thus creating the dream vet horse. Never sound, supported it's whole life. "Breeding vet bills" is what I call it. Culling is a dirty word these days, when it used to be a respected practice." – Bronwyn Margetts

Smooth gaits

are the result of correct work and conformation.

Here the front steps are exactly the same length as the hinds, and there is no gap between the hind and foreleg.

Note the open gullet and roundness of the loin.

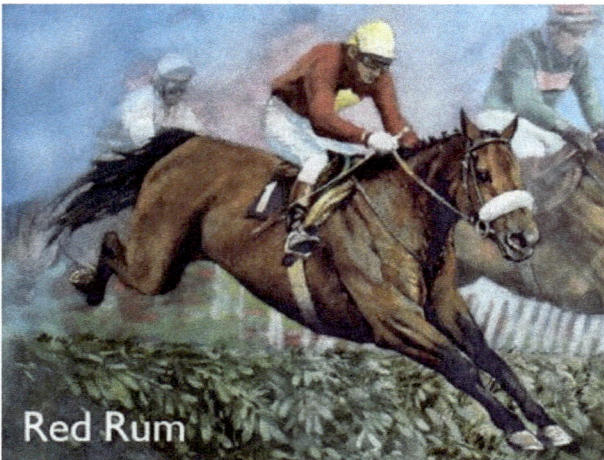

Toughness and purpose, Red Rum Irish Thoroughbred, stands alone in history, winning the Grand National, on no less than 3 occasions. and did not fall in 100 race starts. Look at his depth of girth (stamina), elbow reach and stifle room, two markers of highly athletic horses.

Chapter 11

The right temperament

The biggest question

Is it nature or nurture?

A breeder knows from the foals he produces which temperament characteristics are from the sire and dam if he knows them well. It is then up to him to work with those qualities in his training regime for them to become good horse citizens. They must become useful, either to the breeder or a future buyer. In today's age, where so many riders are not educated, choosing the right temperament is even more important.

There is no such thing as a standard performance test in horses to assess temperament. The closest is the European licensing system for stallions, which has flaws. Therefore, breeders must impose their own standards of excellence.

If you think most of the temperament is in the breeding, breed from the best temperament horses you can find. If you think most of the temperament is how you raise them, become highly skilled in training. It is best to be practiced in both because not all foals turn out with the temperament you want or predict.

"There are no bad horses, only bad handling." While this is true, it is not an excuse to breed from just anything. Don't fall for broad sweeping comments like "all that horse's progeny are like that" or myths like "chestnut mares are fiery." Study the *individual*.

I carefully choose the matings to maximise temperament. Stallions have distinct temperaments and attributes they can pass onto their foals. When you have bred several foals by a stallion, you come to know what these are, even if you don't own the stallion. For example, when I worked at Willomurra, we would know by the behaviour which horse was by which stallion. You could leave a rug on a post in the paddock with Jet Master progeny, and it would still be there in the evening. But if you did so with Warning Flag progeny, the rug would be eaten alive!

The mare's temperament is even more important. After all, she is raising the foal. I will not keep a mare that is hard to catch with a foal on her or one that is aggressive or panicky when you handle her foal. She teaches her foal that humans are bad news.

A mare's behaviour when cycling is also important. Is she dreamy or sleepy when she is in season and grumpy when she is not? Does she show stallion "tendencies?" These can indicate hormonal imbalances or simply be temperament issues. It is best to choose mares that are even-natured, especially under saddle.

With one mare, if I tickled her top lip, she would turn her nose up in Flehman's posture. All her progeny would do this. Another had progeny with a "squeaky" neigh. These characteristics are inherited.

Other behaviours may not be amusing at all but are clearly inherited. If you can't ride horses that buck, don't breed horses that buck! The instinctive tendency to kick out, charge, run away, or strike may be immediately evident in the foal or be noticed years later when the "quiet horse" is unduly frightened. A tendency to jump may only be noticed when there is a huge storm and it is found on the other side of the fence. Or when the owner comes from behind a corner unexpectedly, he is kicked. When frightened, this horse's natural defense mechanism is to kick out rather than swing away or pull back.

A term commonly used in a derogatory manner is "hot." This is for one inclined not to settle (anxiety-related) which can also feed off a tense or nervous rider or handler. A certain sensitivity is required for high level athletes. It is to do with the brain being able to respond quickly to signals and the coordination to follow through. Hot bloods have this ability. But if too sensitive, (and that is relative to the rider, their discipline, and their experience,) a horse can become over-reactive and unsuitable for purpose. Draft breeds tend to have slower reaction times, hence they were known as "cold blooded" and they are not sought out for athletic purposes, but their steadiness may make crosses suitable for police work and nervous adults. Reactions and reactivity are inherited traits and handling brings out the best or worst attributes.

It is also important to notice the difference between similar attributes. For example, ticklishness may be undesirable, but sensitivity may be desired. The balance in breeding for temperament is very much skewed in favour of your personality and training methods, which may not be that of the buyer. Take this into account when choosing your market.

A common mistake of buyers is to assume that behaviour on the ground will be mirrored when ridden. Friendly does not mean a horse will be quiet under saddle. Quiet and friendly certainly helps to sell horses, but it doesn't automatically follow as a ridden horse. For example, a horse that stands still may be "quiet," or have been trained to "stand still." Most buyers will not know the difference and just assume it is quiet. Manners are trained. Ground manners must be taught. The difference between quiet and well-mannered is in the training!

I like my foals born "trained." That means they like having me around, they like being handled, and respond easily to training. They will be curious. I love raising foals, and I love handling them. I like to start them off in the first few days. This requires fewer facilities. It also makes it much easier to send the mare off to stud or if the foal needs veterinary treatment.

The most important attribute of all is *focus*. Can the horse focus on the work at hand, or is he always distracted by the slightest movement, scary object, another horse neighing out, or being away from his mates?

Give your horses exposure by socialising and riding so you know what characteristics are the same as the sire and dam, what the riding qualities are, and what the horse finds easy or hard. What are its responses off the place? Does it settle into a new environment readily? Some qualities, such as courage, attitude, and work ethic, can only be tested under saddle. We all know the "fence-sitter" who has that good one in the back paddock!

What does a good work ethic mean? Most riders would agree that it is a horse with "heart," an indescribable quality resulting from good breeding and training. A horse which, when the chips are down, will press on with good "try." The polo players often told us that the Rannock horses would "rise to the occasion." A willing horse will give it a go, no matter what. Good work ethic means they are tough, bold, smart, know their job, and are "switched on" ready to do it. These horses are likely to be totally bored without a job to do. This can bring on "bad behaviours." Like a sheepdog kept in a city apartment, it is just not suitable.

Here are some words used to describe temperament and behaviours. These words often mean different things to a beginner or a pro. Some are highly ambiguous, and most are subjective. Do you know what they all mean? Which are learned? Which are likely inherited?

Temperaments and Behaviours

Alpha	Girthy	Quirky
Argumentative	Grit	Sensitive
Attitude	Head-shy	Smart
Bombproof	Hot	Snorty
Bold	Impatient	Soft
Cold-backed	Kid-proof	Spooky
Courageous	Kind	Stubborn
Cowy	Marey	Ticklish
Easy-going	Patient	Tough
Fence-walker	Placid	Trainable
Fiery	Pushy	Work-ethic
Gentle	Quiet	Windsucker

Cow sense is highly heritable, yet there is no such thing as a single gene for it. Neither is it something that can be seen in the conformation of the individual. It must be assessed off pedigree and by seeing the horse in action. Some of it is in the mind of the horse, and some of it is in its athleticism. It has to have both.

Courage is most noticeable in the racehorse, and jumper/eventer, yet this is also determined by experience and the rider partnership at the time. Jumpers which dangle their legs will not make the grade, so they must be able to fold the limbs high and "bascule," and be able to shorten and lengthen stride under rider direction if they are to be "clever." The courage, the athleticism and quick responsiveness is inherent, but it must also be nurtured to bring it out.

"It is not enough to find highly talented horses and ride them in an average way. You get better results finding average horses and riding them in a highly talented way." – Jeff Evans, renowned showjumping rider.

Many faulty behaviours are man-made, but I like to avoid them in breeding stock because of any underlying inherited tendency unless I have explored the history of the horse in great depth. Poor behaviours are generally pain induced – teeth, ulcers, kissing spine, pelvic issues, etc. These can bring on windsucking, rearing, head shyness, difficulty catching, girthiness or other unwanted behaviours. If you avoid the behaviour, you avoid any likelihood of bringing the unknown *causal* genetics into your stud. You will experience enough issues over the years without adding to them.

The important thing is to ride your horses before putting them to stud. Know your horses well. The more you have, the less you know about them individually. If you have large numbers, do you really know what you are breeding from regarding their temperaments? King Ranch managed this on a large scale because they only took the best ones for breeding based on the recommendations of their experienced hands.

Temperament is an extremely complex subject. Each horse on the stud is a reflection of *you*, your choices, your preferences, and how you are around them. The most successful horsemen have a calm, quiet "no fuss" approach. Are you that person? Where do you fit?

"Perhaps the greatest kindness you can do any horse is to educate him well "
– Tom Roberts.

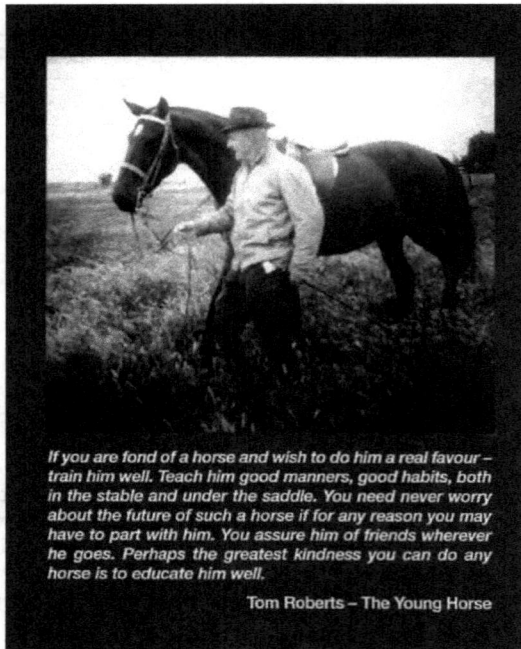

If you are fond of a horse and wish to do him a real favour – train him well. Teach him good manners, good habits, both in the stable and under the saddle. You need never worry about the future of such a horse if for any reason you may have to part with him. You assure him of friends wherever he goes. Perhaps the greatest kindness you can do any horse is to educate him well.

Tom Roberts – The Young Horse

'Tom Roberts, "Go Forward, Dear" by Dr Andrew
Mclean and Nicki Stuart, reprinted with permission.

Chapter 12

Putting it all together

Guest contributor, Peter Haydon

Haydon Horse Stud has an established firm breeding selection regime since 1832, with emphasis on temperament, lightness, mouths, conformation, an athletic hind leg set and action. The emphasis is to breed horses that will perform at the elite level. In a nutshell, the stud looks for a quality horse with an intelligent head, a good correct flat forehead, a sloping shoulder, a light front, straight legs, an outstanding forward stifle set, with a powerful, deep hindquarter. [Mission statement – Ed note].

Selection base

It is a very fine line breeding a performance horse that is trainable, co-operative and who wants to be with you, but at the same time has the character or that special "X Factor" to perform at the highest level. Emphasis is continually put into achieving the right balance between the horse that is just too docile to one that has too much energy, which takes years longer to train and even then, may not make it. The "Champion Horse" has special characteristics and the genetics from these champion horses are widely incorporated in the breeding programme.

Special points

The forehead should be wide, and spirit-level flat, with eyes big and kind, and wide broad nostrils. The jaw should be in proportion and not too heavy or wide, with a fine, clean gullet. A good indication of width is that you can fit your fist in between the jaw bones, allowing plenty of room for breathing and the windpipe.

The neck should be a good length, not too heavy, not ewe shaped, more swan shaped in outline. This automatically puts the horses head in the correct position and does not need the bridle to get it where it should be, it is just naturally there in the right place. There is an old saying that you want the horse to look over the fence not under it. The neck should be set well into a good sloping shoulder with the points of the shoulder up. The more the slope the more a horse can extend its front leg range. Importantly it means the horse can also really walk along.

It is very important the elbows have enough width out from the girth, not being too tight or close. A good guide is that you can fit your fist in the gap. This means the horse is not restricted and saves girth galling. Tight elbows and heavy fronts are undesirable traits.

The forearm needs to be long and the cannon bones short and flat. A horse with shorter cannons can pick their feet up and are not daisy cutters. The pasterns should be not too short or straight and not too long and weak, set at 45 degrees, as they are the horses shock absorbers. The knees need to be straight from the front, not crooked or offset. From the side they should be in a straight line down, not back or calf kneed or overbent. The legs should be straight with the feet following in a perfect line, not turned in or turned out. Deviations should be avoided as well as horses swinging their legs when moving. A straight line is so much stronger when the pressure is applied, eliminating weak spots, helping with soundness and longevity. Feet should be well sized as small feet are less capable of diffusing impact stress.

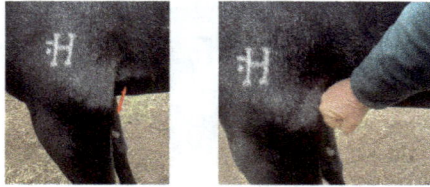

Elbow room

The coupling is very important for athleticism, having enough slack for the horse to move freely, not jammed up and tight. The ribs should be well sprung and rounded, transferring saddle pressure away from the spine and not be slab sided.

The hindquarters are also important and must have the right athletic setting, power and shape to allow the horse to perform at the top level. From the side view the rump should be long and strong with the hocks close to the ground, with short cannons. The hind legs should not be too straight, or sickled hocked with too much bend. The longer the distance from the stifle through the gaskin the better, giving the stifle more power. It is just so important having a forward setting of the stifle which when moving comes well in front of a perpendicular line from the hip. This allows a longer reach of the hind leg, and the longer the overreach of the front foot imprint when walking, the better.

The hind legs should have a straight trajectory when moving. For athletic horses it is deemed preferable for the hocks to be slightly "in" which means the placement of the hind feet will be out a little more, creating more spread and distance between the hind legs when working. This creates a tripod effect which is preferred for stopping and turning at speed. The opposite is a horse that is "out" at the hocks and has a screwy, wobbling action with little ability.

The forward set stifle with its long extension can be seen in many different champions when they are performing at full speed. In fact, it is one of the factors that lets them achieve their legendary status. When moving, the horse with the forward set stifle will have the stifle move well in front of a straight line drawn down from the hip. The further forward the better and the more athletic.

Cuartetera has been rated as one of the best polo horses of all time, playing in the world's fastest and highest rated polo, by the world's best player Adolfo Cambiaso. Her clones have also performed at this elite level. She has an incredibly athletic and powerful hind end as shown in this photo. The top line shows the length from the croup back to below the tail, the next line shows the length and power back from the stifle and the circle shows the forward set stifle. The sloping line shows the length to the hock, which is nice and close to the ground, with a corresponding very short cannon bone. It all means WOW, how athletic!

Cuartetera - rounding of the loins, muscling over the hips, forward stifle. When fitness and athleticism combine. Photo public domain.

In this example of Ellerston Ruski taken as a two-year-old, the right stifle is well in front of the line when cantering and the hind leg is still back, so imagine the reach it has when it comes fully forward. It has the advantage of being naturally set in a forward position to start with.

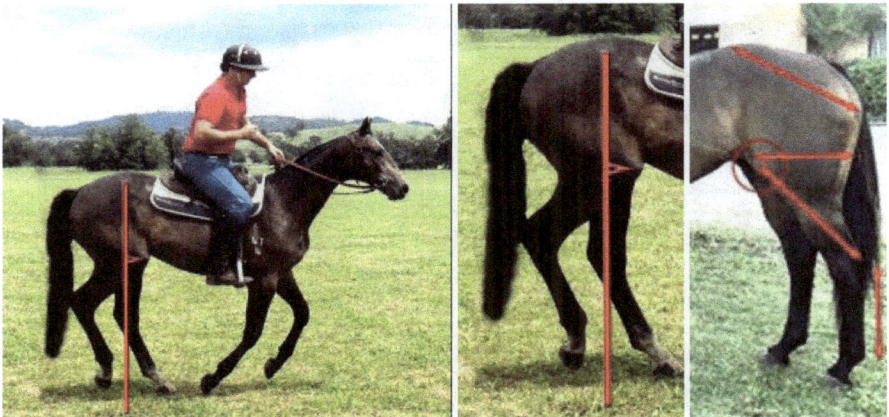

L-Ellerston Ruski R-Cuartetera

Secretariat is one of the all-time great performers, winner of the American Triple Crown and still holding the fastest time in history for all three races. He won the Belmont Stakes by 31 lengths. His stride angle was 110 degrees. Every one degree increase in stride angle means an increase in stride length of 2%. In other words, a 10% increase means 20% more ground is covered.

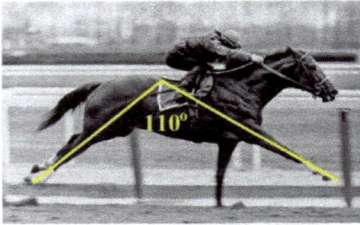

Secretariat measured a stride angle of 110 degrees.

Secretariat- head shot.

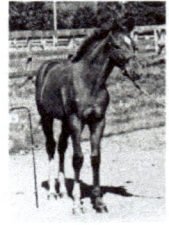

Secretariat as a weanling showing the V shape between the front legs.

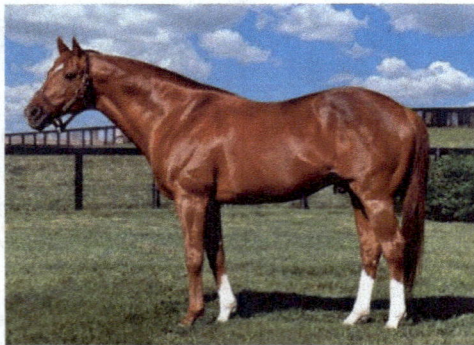

Secretariat was considered to have ideal conformation. Below: Haydon Diamonte

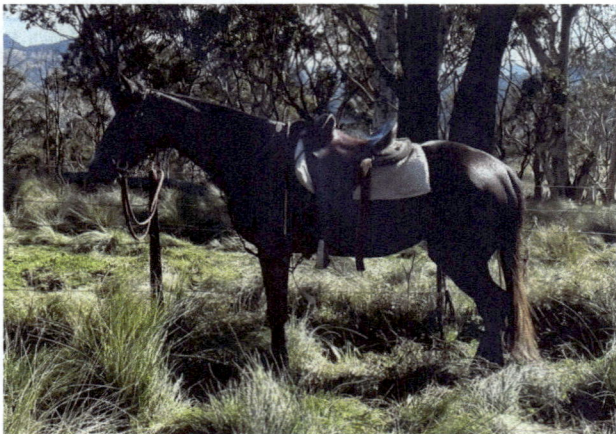

Haydon Diamonte taken here at the top of a mountain range. She climbed easily, navigating well the steep pinches, and gave her rider a comfortable, effortless ride. She has scope and softness. The saddle sits nicely, with the girth well back from the elbows, enabling free movement and no girth galling. The nice length of back softens the ride. Sloping shoulder with the points well up, the light front and length of rein means she can walk briskly all day. The big loose, forward set stifle and rump power gives her plenty of forward propulsion and hind leg reach. Next are examples of great athletes.

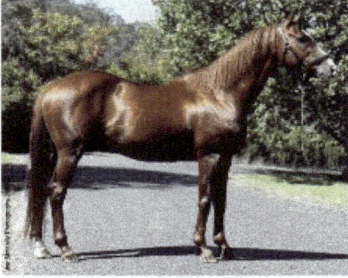

Norman Pentagon one of the world's top polo sires, athletic, sleek, a light front which is nicely up and with hocks close to the ground.

More Than Ready by Southern Halo is the sire of the most winners in history. He has an incredibly powerful, and forward stifle when he moves.

Black Caviar 4 times World Champion Sprinter, undefeated in 25 races including 15 Group Ones.

Winx winner of 37 races, including a world record 25 Group Ones.

Makybe Diva the only horse to win 3 Melbourne Cups

Haydon Angel Jewel won every major polo tournament in the world.

Legendary campdraft mare Breezette

Haydon Eve could plant her hindlegs and do continuous 360-degree spins.

Haydon Dunkirk using his hindlegs in polocrosse.

Wickford Santa Fa showing hindleg strength and power.

Top campdraft mare Hazelwood Romance.

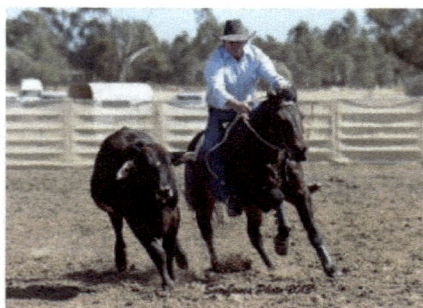

Her sire son Hazelwood Conman showed such natural ability chasing cattle. He is stretching out here to quickly claim the steer.

Chapter 13

The importance of the mare

Buy the best you can afford

"My worst mare is my best mare, or she wouldn't be here." – Howard Pitzer,
legendary Quarter Horse Breeder.

Some years back, when we were in our second generation, we had a very knowledgeable person visit. He was very complimentary about our horses, and asked us what we thought our best mare was. Without hesitation, I replied, "Cat Ballou," and showed her to him. His reaction was very apologetic – "if you don't mind me saying so, she looks quite ordinary against these others." I pointed out that these others were, in fact, her daughters, which were by different stallions and retained by the stud after performing exceptionally well.

She wouldn't still be in the stud if she wasn't producing the goods.

The most important decision you make on your journey will be choosing your mares. You cannot underestimate the value of a good broodmare. A good mare is your greatest asset and the cornerstone of your stud.

No highly credentialled stallion is out of an ordinary mare. The original desert Arabs were all bred on the dam lines, so much so that if another tribe stole a mare, even if they got her back, she was no longer considered pure. Few stallions were kept, and the mares were ridden to war and celebrated. Now with the emphasis on strains, such as the straight Egyptian, it *all* counts. You follow the dam line to establish strain which is evident in the modern horses.

The Friesian people understand the importance of the mare line, and you will often see horses advertised saying 'stam 50'. Stam is the mare line, and the number given to that line can be traced back through the mare books to the original mare which started that line. Various lines are well known for different attributes, and you can read the history to see what they produced.

Not only should a mare be well-performed, but she should be out of a well-performed mare, which is also out of a well-performed mare. This is called the tail female line. In Thoroughbred circles, stakes performers are written in black type, known as "black type" horses. The further back you go until the pedigree is "off the page," you will eventually come to the last female recorded. This is the "tap-root" mare.

If you are breeding in a market that uses black type as its reference, buyers determine what they want to pay for the offspring by the amount of black type in its pedigree, so choose black type pedigrees if you are breeding racehorses. How many winners come nowadays from a pedigree without black type? I would like to know of any in the last 20 years. Too many horses are being bred with fewer buyers, so breed the ones they will buy and give yourself a better chance of hitting the jackpot.

So how much black type do you need? The more, the better, but there are limits, of course. One of those limits is that the black type horses will eventually diminish their ability to pass on the winning genes (law of diminishing return) or will all be related. So new lines, outcrosses, must be found. This is thought to give an element of hybrid vigor and may be quite successful when carefully done.

Of course, there are occasional exceptions, where something good comes from an un-remarkable family, but you cannot count on it in a breeding programme. It happens as infrequently as winning the lottery; that is why they are exceptions.

If credit is given to one key horse in a pedigree, such as the famous Northern Dancer, when others are not mentioned, this is pedigree "licence". If that is the pedigree's only claim to fame, then it says little because almost every Thoroughbred has Northern Dancer in its pedigree. Same with Radium, Abbey, Doc Bar, Three Bars, Skowronek, Justin Morgan. Just pick your breed. By implication, it also says there are no other lines of note.

It is pedigree convention to say that a horse is "by" a certain sire, but "out of" a certain mare. A half sibling is referred to as "by the same sire as" (name a famous horse), never a "half sibling" to it. Only when the half sibling is out of the same *mare* as (famous horse) is it acceptable to use the term "half sibling". This is because a stallion can have 500 or more foals, but a mare likely only 10. Yes, genetically, they are all half siblings, but get this wrong and it is a sure sign of a novice!

Mitochondrial DNA is a portion of the DNA passed down from the mother. Mito-chondria are the cellular components that produce energy. Therefore, mitochondrial DNA could influence some of the finer metabolic and biochemical pathways, thereby influencing performance. In addition, it controls cell differentiation, likely including immunity and the development and progression of some diseases, due to their influence on cell metabolism. The stallion's mitochondrial DNA never enters the embryo and stays outside the egg at fertilization.

Since this part of the DNA is passed on through the tail female line, you should pay attention to a mare's tail female line. I believe the best tail female is a Thoroughbred, for elegance, speed, athletic ability, courage, and many other qualities.

If you have gone back more than three generations, male or female, the influence of that ancestor has been reduced percentage-wise so much that its impact becomes insignificant. The tail female influence is remarkable because it gives insight into those families that continue to influence the present day, quite probably because of mitochondrial DNA.

Simply put, in the early days, those females produced superior progeny all the way to the present. Badly bred stock has tail females, but the names of those are long forgotten, while the good ones remain well-known.

Heart size, in relation to speed, is passed on through the female X chromosome. Stallions with a large heart can only pass it on through their daughters. This is self-evident, as a sire can only receive his Y chromosome from his sire. Therefore, sire sons do not have any genetic material from the sire's X, the maternal chromosome.

Interestingly, the X factor in the Thoroughbred most likely comes from a single British native pony mare, and the breed today that carries more than its fair share of the gene is ...... the Shetland Pony.

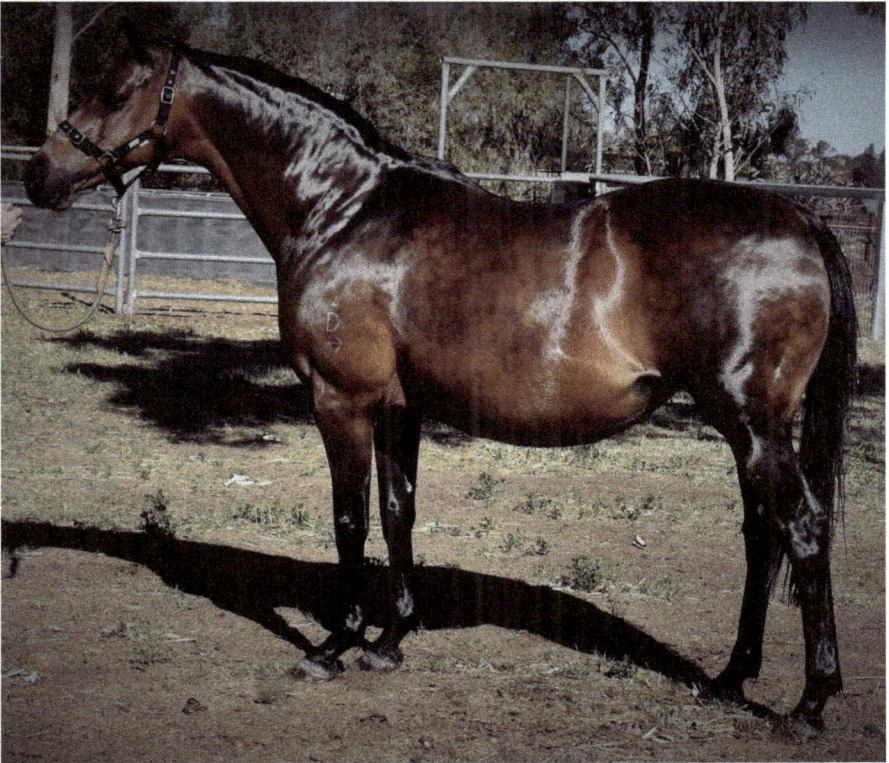

Lanieres, Australian Riding Pony mare. Her dam, Tearna Belle (ASB), known as Cartier won the 1997 Crane Trophy. Owned / Photo by Kylie McKinnon, Kensington Park Stud.

With Australia having some of the greatest Thoroughbreds in the world and the ready availability of Thoroughbred mares for equestrian sports breeding programmes, I cannot think of anything better to use for genetic diversity and athletic ability where the breed allows this. Arabian breeders could use racing Arabians. It is proven time and time again that if you keep speed in your breed, you keep most of the other qualities that you need. Speed is the natural by-product of what nature intended for the horse as its survival mechanism. Speed is the link between form and function.

Most breeders will be reluctant to sell mares from their tail female lines. If so, try to buy a close relation of the mare lines (distinct from the stallion lines). The further back you go in the pedigree before coming to a performed horse on the female side, the more I would dismiss it. Non-performance close on the mare's side is a red flag for me.

If the mare is by a popular stallion, you may struggle to find a suitable stallion to outcross her to, as all the currently popular stallions may be related. However, it is sometimes possible to pick up an outstanding mare from a recognised breeder simply because she is closely related to his best horses, and he cannot use her without close breeding.

Find living relatives of the horses you've liked or owned. Ask the breeder what they have. If you find something, do a deep dive into the mare's story, don't just buy on pedigree or a picture. Find out basics like height, colour, breeding history, progeny, breeding soundness, and get a vet check to confirm all this unless she is currently in foal. Then, look closely at her conformation and temperament, preferably by seeing her in the flesh or a video.

Most breeds have excellent information on their websites. One must research to find what one's breed emphasizes. The Australian Stock Horse (ASH) Society, using information from their website, tells you that HSH after a horse's name, means Heritage Stock Horse. These go back to Waler lines in direct descent without outcrosses to new breeds after the Waler period.

Other breeds list the awards a horse has achieved – Opal Awards, Speed Indexes, or Ratings. These are excellent tools for professional breeders and serious hobby breeders alike. The breeder must research the pedigree and be 100% certain that the female lines he is dealing with are the best he can obtain.

A breed's popularity with pedigree buffs is due to the ease of tracing a pedigree. Unfortunately, this is not possible for some breeds, so it is up to individuals to do their own research and record it. Breed associations with websites that record results, stories, and history enabling breeders to easily and accurately record and update information, have a decided advantage over others. Part of a breed's attraction to newcomers and the long-term converted is the enthusiasm and inspiration that stems from building a profile of a particular horse or bloodline one is researching.

Researching pedigrees and the history of horses can become an obsession, but that is better than not bothering at all. Buyers like you to give them the pedigree and history of a horse, so it is a disincentive if you can't give that information. Buyers like to think that what they are buying is part of something bigger, the community of the breed and its history. You are selling them a dream.

Backyard breeders and fly-by-night breeders don't really care what the dam is as long as it is either the right colour or cheap, thinking that it will give them a foot in the door. As a result, they rarely succeed in breeding anything of merit. They have limited interest in performance, conformation, riding longevity, or even temperament. Rarely will they breed to a stallion worthy of being entire. The resulting foal is worth little in $ value and the breeder quickly realises the cost of breaking it in (and possibly gelding it) is more than he wants to outlay, so tries to sell it on the market unbroken, ungelded, or worse if a filly, putting it in foal!

You are not following sound economics or sound breeding principles if it costs more to break in a horse, or train it, than its eventual value.

The quality of the mare is the difference between you and fellow breeders. Everyone can use the same good quality stallions these days with artificial insemination (AI), but what everyone does with those genetics is different. The quality mare is the differing factor and gives you the competitive edge. I don't think breeders "forget" the mare; I just think many use sub-standard mares. Good broodmares are harder to secure in their prime, and it takes several years for a mare's progeny to prove themselves. It is easier to promote the stallion with glamorous photos and show wins. It is easy to talk about his percentages of successful progeny, which makes it easy to overlook the numbers that weren't successful.

Understandably, stallions are the main marketing method for a breeding program. *Yet the mares are the more influential half of the cross!*

When you are more experienced, you will find owners offer their good mares to you for lease or private sale before they advertise. This gives you time to explore their suitability and to obtain something which would not otherwise be possible. The top breeders will often do swaps and decide with a handshake agreement. Others will give considerable discounts. They can because they know you will do right by them.

The importance of the mare. Sandhaven ASH stud, Sierra (granddam 22 yr old), Mischief (dam) and Sandhaven Late to the Party. Owner and photo, Kellee Campbell.

HB Sezuantino, Hanoverian Pre-Licence colt, 17mths, owned / photo by Amanda Evins. Below: Regal Banquet, Thoroughbred stallion, with Jayne Anderson, owned by the Powell and Anderson families. Photo Foxwood Photography.

Chapter 14

The stallion

Do you really need one?

"The best stallion is the one in the freezer." – Kellee Campbell, Texas vet and horse breeder.

Most studs will never need a stallion. With the availability of AI, unless you have at least five breeding mares, it is simply a convenience and false economy. Never keep a stallion because you might profit from his service fees, or because it is cheaper than sending your mares off site. (In the long run it's not, as we shall see!)

Any decision to own a stallion comes with added responsibilities. Horses are inherently dangerous, but stallions more so. A stallion, no matter how quiet it is, is still a stallion, and quite capable of changing from a pussy cat into a dangerous beast in an instant. He must *never* be taken for granted. Other people may do stupid things around him, especially children, or novice riders, which you need to watch for. There may be a loose horse on the grounds. He should always be treated like a normal horse, *but with the heightened awareness that he is a stallion.*

Any decision to own a stallion must come with considerable forethought.

- Can you afford to buy one of excellence?

- What is your experience handling a stallion?

- Who is going to handle him for stud duties?

- Who will ride him?

- What facilities do you have to house him in a suitable manner?

- Can you afford the extra upkeep and time involved?

- Can you take on the responsibilities of outside mares?

- How will you promote him?

- What extra costs are involved?

- If you don't take outside mares (a closed stud), are you willing to do all the promotion yourself?

Any breed only needs one stallion per 30 mares in the studbook to retain good numbers and good type. Rare breeds have difficulty getting close. Volume breeds are way over this, as every owner keeps a stallion in case it becomes the next "best thing." The low-end market is awash with poorly bred stallions with no performance to recommend them. This gives all breeders a bad name.

I would delay the purchase of a stallion until you need one, at least a generation down the track. This allows you to become sufficiently experienced to understand what lines to choose and where to look for the right one. It may even be a colt you breed yourself. Sometimes one comes along that is too good to geld and too good to sell. There is no reason why you can't run it on to see how it develops.

Most worthy stallions will be expensive, so you need to be highly selective. The horse must satisfy higher criteria than when selecting the mare. Though he doesn't have as much influence as the mare on *individual* foals, he can make or break you if you choose the wrong one. He can sire a lot of foals in a short time. They must be above average before he is likely to make the grade. Will you be able to sell his progeny for the price you would get if you chose an equivalent, or better, outside stallion? A stud's reputation is usually built on its stallion, good or otherwise. Unless he is promoted exceptionally in his first few years, it may take 100 progeny before buyers and breeders take serious notice of him, and that may take a lifetime, too late for most stallions to be recognised. Standing a stallion is not a decision to take lightly.

Mare owners will come to you with many questions, varied experiences, and knowledge. Much more paperwork is involved – contracts, service certificates, breeding returns, and the like, depending on your breed requirements. You have just increased the size of your business. Are you a people person? Can you conduct business in a professional manner?

How will you campaign him? This takes a lot of thought. A stallion must prove his worth and come before the public to create awareness. Don't think mare owners will beat a path to your door because you own a stallion of the greatest bloodlines. They must know about him, and they will want photos and recommendations.

Can you adjust to the requirements of the breeding season? The time it takes out of training or competing with your other horses will be significant. During the off-season, where will you house him to give him a chance to be a "normal" horse? There is no reason why you can't put him in with an in-foal mare or a suitable gelding, and some studs may even run a stallion with other colts.

When looking at a potential stallion prospect, the first question is, *"will he make a good quality gelding?"* If the answer is no, then walk away. He is not good enough to remain entire.

Are you buying off a photo? Remember, photos can be photo-shopped, so ask for a video. You want to see both on-farm for movement and off-farm for manners and trainability. Who is handling him?

Chalani Nightdance HSH stallion owned by Ashborn Stud. Photo by Kim Ide.

You may choose to buy a young colt. The downside is that you must do all the promotional work yourself and wait until his progeny are old enough to evaluate. If you decide they are not up to scratch, you need to start again. In the meantime, he will likely compete against other colts of the same line, so the only difference will be the mare line he represents and his performance. Make sure he has a reputable mare line.

If you buy or run a colt yourself, the best practice is to raise him with an in-foal mare or a gelding, to teach him socialisation skills. Or run a few colts together. But it doesn't teach them to behave suitably around mares, so I prefer it to be an in-foal mare.

An ideal way to purchase your first stallion is to find an older successful horse with a record on the board, both under saddle and producing the type of progeny you like. It is important that he has progeny to view, and you can see the mares they are out of. Some stallions will only breed better than themselves if the mares are good. Others will upgrade an average mare but not good mares. Some stallions don't appear to pass on their faults.

What good qualities does he have that he is passing on regularly, regardless of the mare? Are these the exact qualities you require to improve your herd? Will he be an outcross to your lines?

People sell stallions for a variety of reasons, one being they are keeping his daughters. Check if these are consistent as to his good qualities and type (ignoring colour as that might vary). What is their trainability or performance record?

I would never buy a stallion without a suitable temperament. This means the temperament you seek in his foals, not just in him as an individual. How he behaves when he is out and about is how people will judge him. Does he have manners? People often excuse a stallion's behaviour, but that doesn't mean he should be bred. Is that what you'd want to ride? I would question using a stallion that hasn't proven himself "off-farm" in some way. This would come down to the performance of his progeny.

Fames Presence, Arabian stallion owned by Carolyn and Margaret Potts. Photo Rob Hess.

Standing at stud

If you own a stallion, you can offer him at public stud, by private treaty, or keep him only for your own purposes, whereby his book is "closed." A private treaty simply means for someone to contact you; you may do a deal, depending on the mare. However, if at public stud you should still be selective about your mares. For example, you might decide you won't accept mares of too close breeding, the wrong genetics or colour, or difficult breeders. But importantly, you should reserve the right to reject mares of doubtful temperament or conformation. This should be in your contract. Examples of stallion contracts are on many breeder's websites, which you can download and rewrite to use for yourself.

"Walk-in, walk-out" is ideal for the smaller stud, with limited facilities, as it puts the onus on the mare owner to work with their vet and time the breeding. When the mare is "right," the owner brings the mare across to be bred and takes the mare away again. This method has plenty of flexibility, as you can keep the mare overnight or even for the week till she goes "off."

Will you be asking for payment upfront or a deposit first and payment upon a positive pregnancy test at 14/45 days? Will you be offering a live foal return, and under what circumstances? We like to do so because we really want our mare owners to have a foal. This simply means that you allow the mare owner to return the mare (or a substitute mare) the same or the following season if a live foal is not produced. Whatever you choose, make sure you have a contract to cover your circumstances. Include whether the mare needs a clean bill of health before she comes, which saves you and the mare owner a lot of time, or whether you will do that upon her arrival.

There are some contracts with a "no-colt" clause, or different fees, depending on the sex produced. These are hard to enforce, and may be breaching Trade Practice. Better to have incentives stating that if the foal is gelded, the mare can be returned the following season at a discount. If the mare is of insufficient standard for an owner to keep a potential stud colt, why accept the mare in the first place? I would not use a stallion with such a contract.

Quarter Horse sire Jet Master AAA, owned by Willomurra - Photo by Peter Gower.

Stolen Identity, Appaloosa stallion, Stolen Identity partnership. Photo by Gail Smith.

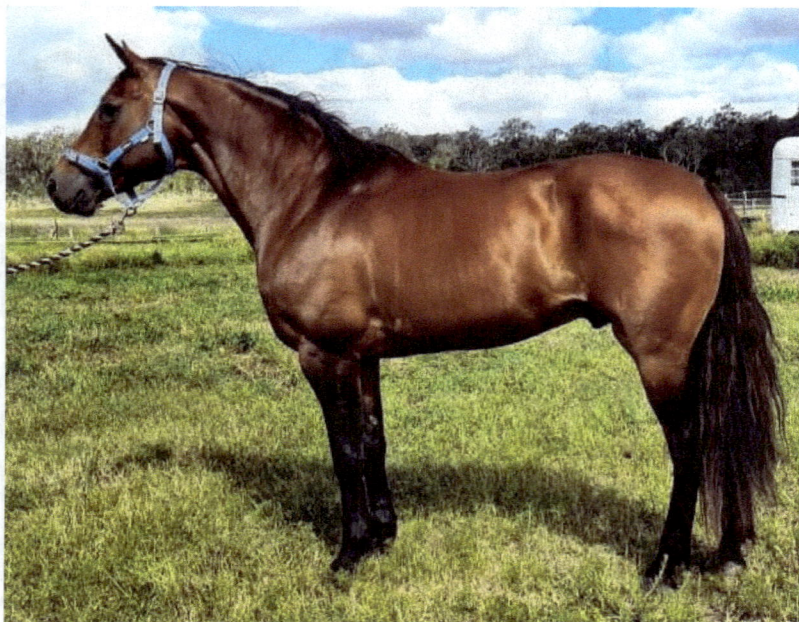

Warratah Aurora Australis, Waler stallion. Owned/photo by Penny Bieber.

Chapter 15

The next stages

How to decide which foals to keep

Quality is remembered long after the price is forgotten.

It is said that the best time to evaluate a youngster is at three days, three weeks, three months, and three years! Another time to take a good hard look is after some rain when the foal's fluffy coat is lying down. You will notice in the second photo that the foal coat is not completely shed. It is easy to see that the baby is well balanced, has a wonderful shoulder, good gullet, stands square, and has a nice hindquarter.

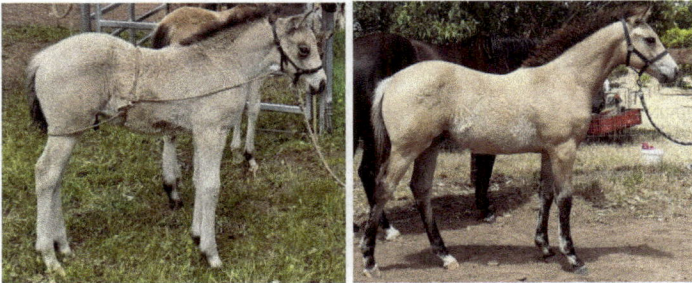

Chalani Yellow River ASH at 3 weeks and 3 months.

All foals lengthen as they get older because at birth their legs are longer proportionally to their bodies . So the perfectly balanced-looking chunky foal may look longer barrelled and short in the legs at maturity. The longer-legged foal may take a while to reach full height yet end up very well balanced overall. The withers are the last bones to mature, but you can tell at a young age if it is going to have a decent wither by feeling the top of the shoulder. It will feel flat, or "blocky" if the withers are poor.

Baroque breeds are slow maturing, and the youngster will look nothing like an adult, with its large head, no topline, and weak hindquarters. These horses are bred for artistic pursuits, such as dressage, and with that education, their bodies develop strongly and they become increasingly graceful and elegant.

Sometimes a decision to retain a foal may be obvious. She is the last of a certain mare, or a colt strikes you as the standout. Perhaps it is the one by an outside sire? If you have a range of good fillies in one year, how do you decide which to keep? Should it be the one with the best conformation, character, or bloodline? Follow the principles for the selection of your adults. Perhaps retain a few until they are much older, possibly broken in. This should add value to the ones you choose to sell.

You can usually tell the approximate mature height of a youngster by comparing the height of its hocks with that of the mare. If it will mature the same height as the mare, the top of its hocks will be nearly the height of the mare's hocks. Some people use the "string method," which measures the elbow to ergot, then the string is flipped taut from elbow to wither. This is the estimated final height. It is most accurate at around 18 months of age.

Some lines have nice heads, which plain off considerably as they age. Some shorten in the neck as they grow, and others lengthen. Some foals are striking movers but not as adults. Others are not balanced enough to move well until maturity. If you are a buyer, you must judge a foal by its parents and siblings. The rest will be your faith in the breeder's expertise in knowing his lines.

The next phase

After a few years, you will start to feel established. You still have to evaluate your stock every few years or at least every generation. Which horses are delivering the goods? Maybe they need to be retired or sold. Start consolidating your breeding stock to keep only the best of your best. In this phase, you won't want to diversify too much. Instead, stick to the lines that are working for you and develop them further. Don't chase too many out-crosses at once. You can undo everything very quickly. This will not seem evident until your first generation is on the ground.

As you progress you need to focus on making fewer avoidable errors as opposed to breeding spectacular winners, by continuing with proven trusted methods. The best way to explain this is that during booms it's easy to look smart. Rising numbers of buyers cover up many mistakes in the industry. A good economy assists all horse breeders, masking the worst. When the economy is stalled, it reveals who is unable to do the right thing, who is not reputable, who is using poor methods, who can no longer cope. Warren Buffett explains it well, in two of his famous sayings:

"A rising tide lifts all ships"

"When the tide is out you see who is swimming naked."

Horse markets around Australia are very fragmented and after a period of reasonable growth, seasonal or economic factors bring boom times to a quick end. Historically there have been many reasons why people have gone out of horse breeding (a good thing) or horse buying (not so good) and these include more recently Covid, drought, fire and flood.

The unfortunate truth of owning and running a business

"Running a business is really hard. What they don't tell you is that it can cause severe stress and anxiety, and drains you mentally to the point of depression in even the most laid back people. People will talk about you, compare you to others, use you, they will view you as a

service and not a person anymore. Friends and family will expect discounts and people will value you and your hard work less than a big chain store.

You have to worry about if you forget to email/message someone back, are they going to think it was on purpose? Did you disappoint them? Will they hold that against you? When in reality you just can't get to everyone's messages and emails. Starting up and running a successful business puts incredible strain on personal lives and relationships, many of which fail because there is no work-life balance. You need to be the director, the worker, the admin, the marketing team, the accountant, the cleaner..... All whilst being a parent, a husband or a wife, family support, friend... it's one of the hardest things you will try and balance.

There's a reason you don't see many people succeed in small businesses after 5 years. If they are successful they are overwhelmed. It takes a toll. It's freaking exhausting. Especially the past couple of years when so much has been out of our control. Here's a small reminder that we are just normal people with hectic lives. Be kind, be patient, support small businesses....... and hopefully more will stick around!" – author unknown.

Some horse breeders survive year after year producing quality, useful stock and have a steady market through word of mouth and repeat buyers. These successful breeders are not focussing on trying to get rich, but to maintain consistency. This is the time to re-examine what you are doing, the why, and to reset your Mission Statement and goals as needed. You will need to do this with each generation.

It is said not to have rose-coloured glasses. However, all the top breeders tend to be over-critical of their horses because they know them so well and compare them against their ideals. They are certain of their ideal, always breeding to raise the bar. If you are not like this, you are probably not breeding to a high enough standard. Emotionally, this is the hardest stage because you have become attached to your homebreds. This phase, in my opinion, is where most breeders fail.

Hoping you will breed better horses, hoping you will move forward by keeping them all, is a fantasy.

The next generations

By now, you have realised that you don't make money breeding horses and that it is very hard to breed and compete simultaneously. So you may be better off buying instead of breeding if you'd really prefer to compete. But the reality is, you can't stop competing with your homebreds if you want to promote your horses and stay in business *long term*.

You are now an established stud with a great line of quality mares. Where do you find that next stallion? The risk of using something that introduces faults you never had is now *high*. Your selection must now focus on maintaining the good genetics you have solidified in your programme. How do you improve one thing without losing some of your other attributes? How do you find that horse which will take you to the next level?

There is no one answer to this dilemma. It is a good dilemma to have. Establishing a line of quality mares is harder and takes longer, so keeping your females and changing your stallion is better. It requires a careful search, possibly using international lines, a new upcoming line, or a nearly forgotten old line. Once you are an established stud, you become a trendsetter if you choose well. If not, you can undo everything very quickly.

This may be the time to have two stallions. A chance opportunity too good to pass might arise at the right time (or even before you are quite ready). However, I can never understand the philosophy of some studs having multiple stallions, especially of several breeds, colours, or registrations. Keeping up to date with memberships and the paperwork is in itself a huge task. The belief, it appears, is that you will cater to the taste of all mare owners and buyers. This is a false economy. One cannot possibly do justice to them all. These studs become known as the mass producers of average horses. They tend to disappear after a few years and dump horses onto the market with little thought about their future.

A breeding vision takes many years to evolve. Breeding en-mass will not make it happen any quicker.

"You cannot make a baby in a month by getting nine women pregnant." –
Warren Buffett

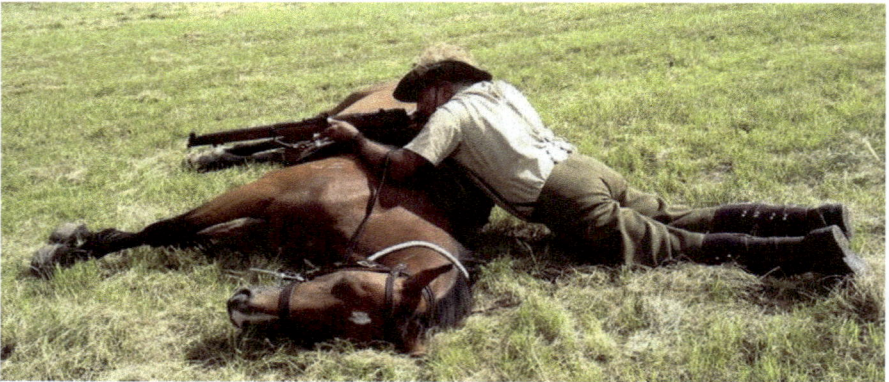

Emu Gully Gunfire HSH wth Ronnie Roman, doing light horse re-enactment. They also do parades, and equine assisted learning. Photo Lynda Rodgers.

Charge at Beersheba re-enactment

Chapter 16

Artificial breeding

The decision to breed by artificial insemination may be breed dependant. Sometimes it is for convenience. Other times it may be necessary due to distance. Or the mare may need to be at a reproductive clinic to enhance her chances of getting in foal, due to a poor history. You can do stallion collections yourself if you have the facilities. You can even become accredited as an Artificial Breeding Technician following the Australian Veterinary Practice Regulations.

Offering artificial insemination instead of live cover may be more realistic for some breeders. Some stallions have never served a mare naturally, usually those out competing, while the Thoroughbred racing industry will not allow artificial breeding of any type.

Fresh semen can be collected and inseminated into mares on-site or nearby. The stallion is led to a phantom or dummy, which he mounts and then ejaculates into an artificial vagina. Different stallions may prefer different AV types, temperatures, and pressure. Semen is analysed for concentration and motility/morphology and then processed. The amount and quality of harvested semen varies between stallions.

Chilled semen must be collected and transported within the time frame required by the off-site mare owner, sometimes hundreds of miles away. It must be timed as close as possible to the mare's ovulation, and chilled semen is generally only viable for 24-48 hours. This makes the logistics of travelling semen, with the right connections between planes and travel companies difficult, even when all goes to plan. It means liaising with a trusted reproductive vet on both ends. You may need an available in-season mare to make the collection or take him off-site to a collection facility. Timing is critical. If you have a job, taking time off work may be tricky. If the mare doesn't get pregnant the first time, you must do it all again.

Frozen semen can remove some of these difficulties for the stallion owner. Arrange beforehand by deciding how much you wish to store and whether you will collect for the international market. Choose a facility set up to quarantine per AQIS requirements and one knowledgeable in each country's quarantine restrictions and requirements. Pregnancy rates for frozen are not as high as chilled, but the gap is closing rapidly due to advances in technology and the experience of vets. It is an excellent way for the small breeder to maximise his stallion usage. The mare owner can purchase it ahead of time and use it when needed. While frozen is great insurance for your stallion, there is the possibility you may never use it yourself.

Semen has been used up to 20 years past collection, and the time frame is increasing. It is also possible to collect semen from a deceased stallion if the testicles are removed immediately and kept chilled in an insulated container. Contact your vet asap if you need this procedure. Don't forget to factor in annual storage fees.

Some stallions don't freeze very well. You might need to try different extenders and freezing protocols before finding the ideal for your stallion. Once frozen, you, as the stallion owner, merely sign a release, and the clinic manages all liaison with the mare owner's vet, including thawing procedures. Basically, (refer to contract) you have no further input, which makes it very convenient.

The main disadvantage of AI is that you have limited control over the suitability of the mare coming to the stallion, or the mare owner is unlikely to have viewed the stallion.

If you decide to import semen, try to go through a reputable agency that can guarantee the quality of the semen. If dealing directly with the owner, you have to go by their knowledge or the information they are willing to provide. Most frozen use is in the Warmblood and rare breeds sector, where frozen semen is regularly imported from Europe. Unfortunately, some is poor quality, so buyer beware. If this is the case, you may need to look at extra straws to get the mare in foal.

Buying by the dose is the most common way. The laboratory that freezes the semen will recommend the dose. Less commonly, some stallion owners will sell by the straw, where the buyer nominates the number of straws they want to buy. Most mare owners don't know enough about semen to judge how many straws they need, and their vet should really decide that based on the technique they will be using and the lab report. Also, factor in the hire/purchase of a storage tank and freight charges. Companies like Gene Movers will oversee the transport process. With AI, each attempt will have a collection and processing fee.

The lab report should give you the post-thaw quality of the semen – sperm concentration, its morphology, motility, progressive motility (after 30 minutes), and, if possible, a viability rating. Check if a 24-hour culture incubation and count was done on the sample. Pass the results to your vet to ascertain if the semen is worth purchasing and how many straws you need. If it turns out the dose has a high bacterial count, you may not get the mare pregnant, and worse, she may need to be treated for some time to remove the infection. Anybody who has ever had a mare with a yeast infection knows how difficult this can be to treat.

There is an excellent series for mare owners on understanding the freezing and thawing process in this series of videos available on YouTube. Frozen semen for mare owners: https://www.youtube.com/watch?v=MeZIVpOUwY4

In the future, intracytoplasmic sperm injection (ICSI) may be a more common use for semen. This procedure requires only one sperm cell to be injected into an egg and, therefore, would conserve sperm banks. When used, it produces the same pregnancy rates as normally frozen insemination. Currently, the cost and lack of facilities that offer it are limiting.

It is also possible to buy "sexed" semen, but it has not yet advanced to the point where it is reliable or inexpensive, and the facilities that offer it are limited.

Remember, if you spend a lot of money using all the "latest techniques" it doesn't necessarily make the resulting foal more valuable in the the long run. *You may be throwing money away. Consider your options very carefully.*

Embryo transfer

This technique is becoming more popular than ever due to reduced costs. It is used mostly to allow good quality mares to continue performing or mares that have problems carrying a foal to term. An embryo is flushed from the donor mare at 7-8 days and recovered by filtration. Transfer can be into several surrogates over a season. Check with your Breed Registry how many are allowed to be registered per year from one donor.

It is possible to transport embryos for short periods of up to 24 hours at reduced temperatures, and after freezing, they can be maintained for very long periods. Embryo transfer success rates are now 50-70%.

The procedure requires a surrogate or recipient mare, which is often not carefully selected apart from having a uterus. She may be difficult to handle, which is not in your best interests, as you are responsible for her care while on "lease" from the company which provides her. Usually, you cannot use a recipient mare of your own. The surrogate must be timed to ovulate simultaneously with the donor mare. Check whether there is a live foal guarantee and when the mare must be returned. Costs are always reducing, but it is still not worth doing except with more valuable mares. Some recommend supplementation of the recipient with omega-e fatty acids to support gestation.

Cloning

Cloning, while expensive, appears to have a bright future where a Breed Registry allows it. Most won't because it is impossible to know by DNA testing, which of the clones is the sire or dam, as all will have the same DNA as the original. Cloning has been used considerably in the polo field, and with some show jumping/dressage geldings to produce colts for use as stallions. It plays an important role in genetic preservation.

Cloning has demonstrated that white markings are variable between clones, even though they are genetic replicas. This supports the theory that genes may only provide the switch on/off for various functions, and developmental factors in utero play a more critical role. In time we will know what other information is supported by developmental factors rather than genetics.

Cloning is done from tissue samples taken from stem cells or skin cells. If you are looking at cloning a deceased horse in the future, make sure you collect tissue samples from the live horse! Speak to your vet about this. The cost of preservation and storage of the cells is not cheap, and the actual cloning process is high. Although it is becoming less due to technology and competition, you could still be looking at $100k per foal produced.

You can view this website for some very informative videos and articles beyond the scope of this book: Equine Repro.com LLC: https://equine-reproduction.com/

Now mares and stallions which are problem breeders, for various reasons, are able to produce foals with the help of modern science. As breeders, we are guardians of the breed, and I feel we are setting ourselves up for more problems. Just because the vets have the tools to keep the unsound horses sound, doesn't mean breeders should breed with them to produce more unsound horses. Just because vets have tools to breed from the mare "difficult to get in foal" for various anatomical or veterinary reasons, doesn't mean it should be bred. *Consider your choices wisely.*

Berragoon Hallie HSH ridden by Lucy Grills in 2016. Dam of 20 registered foals (ET).
Photo Joe McInally.

Chapter 17

Breeding methods

You can't be a successful breeder without understanding the laws of inheritance. Even so, you will sometimes get surprises, not always for the best. You may also find that every foal differs when repeating the mating several times. It is a chance combination of both the sire and the dam. The dam and sire got everything they are from their dams and sires. And so on and so on. Hence look for consistency on both sides of a horse's pedigree.

Always ask yourself, what would I want to ride? Can I breed a horse that makes my job as a rider and a trainer easier? Am I breeding for commercial avenues, or do I think the cross will be a good animal for me to continue with? Can I breed something as physically sound as possible to avoid performance issues later?

A good pedigree means nothing if the horse with the pedigree is lacking in structure, temperament, or movement. It has "lost" the good genes in the pedigree and, therefore, cannot pass them on. Don't breed a less than average mare to a top stallion. The stallion only provides half the genetics. Even if the result is reasonable, the next generation will likely not be.

You never get the average of both parents or a blend of the two. For example, if one parent has turned out front feet and the other has turned in, that doesn't mean the resulting foal will have straight legs. *Look at the parents' worst traits to see if you could live with, or market the foal, if it received the worst of both parents.*

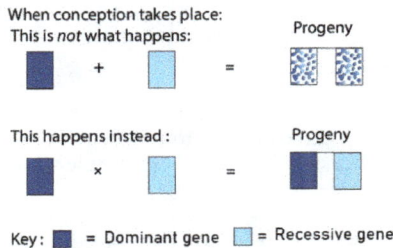

When conception takes place:
This is *not* what happens: Progeny

◼ + ◻ = ▨ ▨

This happens instead :   Progeny

◼ × ◻ = ◼ ◻

Key : ◼ = Dominant gene   ◻ = Recessive gene

Similarly, in cross-breeding, you rarely get the best or even the "average" of the two breeds. There can be much variation between the crosses, even with full siblings. Cross-breeding is best only between similar lines, strains or breeds.

All the foal's genes come from its parents and their parents. The term "throwback" means a certain individual resembles a past ancestor. The idea of "throwback" tends to suggest that for no reason, certain genes that weren't there popped out suddenly, as it were, to reappear in the form of that ancestor, but this cannot be the case. Except in the form of a rare mutation, nothing that wasn't there to begin with, can appear. All it means is that existing genes in the parents are reassembled into a form that resembles the ancestor.

In fact, it might only be in one area it resembles the ancestor (like colour) though most of its other characteristics don't. So, one cannot assume that because a horse resembles an ancestor, it is like it in other features such as hardiness, ability, and temperament, or that it will reproduce like the ancestor.

Inbreeding or outcrossing?

Breeders have a lot of discussions about whether one should use inbreeding or outcrossing to achieve the best results. Very bad breeding is where one repeats a mating that doesn't work. Good breeding is where something well-planned, using quality individuals, goes really well. It depends on the individuals used. Good or bad can come from either method. *Inbreeding is merely a tool for the breeder to concentrate good genes, and outcrossing is merely a tool to diversify and find new blood.* The best horses are generally inbred, but the best breeders use both methods.

A valuable breeding method is to use "nicks." This is where one line, crossed with another line, is known to produce good results repeatedly, often better than both parents. For example, we knew the Rannock line crossed particularly well with the Terlings line and vice versa, so we repeated this. We also liked the Pipe of Peace (TB) line crossed with Midstream (TB), so we purchased the stallion Splashdance, a cross of both, which proved successful for us.

If you find a nick has been successful with another breeder, copy it. Don't be afraid to copy what other breeders are doing if that is the type you want. There are many well-known nicks in most breeds. You may even discover a nick of your own. Once you are into your second and third generations, crossing your own bloodlines together can create excellent "nicks."

When an inbred horse is outcrossed, then bred back to an inbred horse, then bred out again, and so on in each further generation, it is called "reciprocal backcrossing."

Thoroughbred breeders have always had a dilemma. To breed a true stayer (say two miler), you would imagine that breeding the best two miler to the best two miler will produce a champion stayer. Then after a successful racing career, breed the progeny of that to another champion stayer. Right?

No, it doesn't seem to work that way. The more you breed stayer to stayer, the more likely you are to get a horse fast over three miles or four miles, but not the type of stayer fast over two miles. In gaining distance, you lose speed - (a steeple winner, maybe). Do that again; you will likely lose more speed and get an eventer!

Looking at it from the opposite direction, fancy breeding a sprinter? Breed the fastest to the fastest, right? No, keep doing this, and over generations, you lose distance. Your Thoroughbred eventually may only be able to sprint over Quarter Horse distances. Breed those together, and eventually, you may lose even that distance and only breed fast roping horses! This phenomenon is known as "generational loss".

What do experienced Thoroughbred breeders do about this? They will sometimes judiciously breed their stayer to a sprinter, sacrificing one generation to get something else. They may be successful in breeding a miler. They may be successful in breeding a beautifully bred also-ran. They could even breed themselves a winning racehorse, though not a champion. But then, they keep the progeny and breed it back to the stayer. Presto, they are likely to have a true stayer with a turn of speed. On the other hand, a great sprinter may be bred to something of a longer distance, and then the process is repeated.

So this is the principle of reciprocal backcrossing - the art of planned breeding back to something that ordinarily appears to be unexpected, for a long-term gain. I stress long-term. It requires a vision, a goal, and quite an element of risk.

As evidence, you will often hear the debate as to whether Thoroughbred blood should be introduced into certain breeds where they are allowed. When we bought our Thoroughbred /ASH stallion Splashdance some years back, we received a lot of criticism for using a Thoroughbred on our Australian Stock Horse mares; others said we were "forward thinking." Of course, the plan was to introduce qualities that fit *our* mares, but the plan was also long-term, to cross progeny back to heritage stallions. In another generation, we may choose another Thoroughbred stallion to cross back over them, once again using the tool of reciprocal backcrossing. This way, we should be able to keep the speed in our polo pony lines and improve quality, but keep the Australian Stock Horse type we love.

If you are buying a horse specifically as a future breeder, it is valuable to select an inbred individual. If you are selecting something for a performance prospect along with its breeding value, you might look for a horse resulting from reciprocal backcrossing.

Inbred horses are more consistent producers of desirable traits due to increased "homozygosity" or doubling up of their desirable genetics. In other words, they get their good characteristics from both parents. Outcrossing can produce the same result but less often. When you look at the breeding of most of the best-performed horses, you will find noted ancestors repeated in their backgrounds. Although this is called linebreeding by some, *all linebreeding is inbreeding*.

The danger is that as breeds specialise more heavily into disciplines and select for performance on pedigree and results, certain lines and even breeds will bottle-neck. Genetic diversity is lost, and horses become that type, whether that type is deliberately selected or not. Most rarer breeds, or breeds with geographical isolation, have already reached this scenario, so rules have been established to prevent inbreeding, even though by outcrossing, they may not be using the "best" horses. Once numbers build up, it might be possible to allow more breeding freedoms. Zoos are facing the same problem with their endangered species. The use of frozen semen from all over the world and shuttle stallions engenders outcrossing. Still, bottle-necking will occur eventually when the most prominent rise to fame in the next generation.

> *"Market-based selective breeding, which for the modern American Thoroughbred means seeking to produce yearlings that will sell at auction for the highest prices, increases potential for the disproportionate representation of popular sires."* – Dr Deb Bennett, PhD.

In breeds where it is mandated not to inbreed for, say, the first three generations, it is nearly impossible for most traits to be prepotent. Thus, every foal can be quite different from full siblings. Also, parents may be great themselves but fail to pass that combination of traits to offspring. As a result, the offspring typically is a random mix, and you may have no idea where the traits came from because neither parent has what the foal has. You try to understand the horses in the pedigree three to six generations ago to predict what might happen with each mating.

All breeders need to be aware of these difficulties in their breeding programs and their breed. So looking at "hypo-matings", or the projected pedigree of future foals, is a must. Tesio, one of the most successful breeders of all time, used to keep a book with each page split in half, with the pedigrees of stallions at the top and the dam's pedigrees at the bottom. This allowed him to make decisions based on the extended pedigree of the future foal. For example, his greatest triumph was to breed Nearco. If you examine Nearco's pedigree there appears to be no inbreeding in the first three generations, however, when you come to the fourth there is St Simon, top and bottom, as well as in the fifth.

Some would call this line breeding, when it is after the third generation, but it is all a form of inbreeding. It must be both on the sire and dam's side, otherwise if it is only on say, the sire's side, that is the *sire* that is inbred, not the horse in question. When an ancestor is on both sides of the pedigree in roughly equal proportions, that is known as "mirroring."

"A stallion passes his Y-chromosome exclusively on to his sons and his X-chromosomes exclusively to his daughters. This could explain why line-breeding to a stallion via a son and a daughter is exceptionally successful. I prefer line breeding from the fourth generation because this makes it possible to apply powerful line-breeding while preserving sufficient free generations that ensure variation of the DNA. At the same time line-breeding from the fourth generation ensures that the first three generations are free of line-breeding , which also ensures variation.

"The most powerful instruments available to a breeder for successful results are in-breeding and line-breeding. These allow him to connect bloodlines. You can see it as the mortar used for the construction of a house, and it ensures the stallions that are successively used in the creation of the individual are not stacked up like loose bricks. Too much mortar will result in concrete and unskilful bricklaying in bad quality. Balance is therefore the key word for the effective use of in-breeding and line-breeding. And when that balance is achieved, the relevant horse has a higher chance of developing into an exceptional showjumping horse. Such a horse has a better chance to make an indelible mark and to turn out to be a corner-stone."
– Jac Remijnse, Warmblood breeder.

How much is too much inbreeding? In laboratory mice, it is about 20 generations, crossing each generation back to itself. Fertility and vigour suffer greatly. This is known as "inbreeding depression".

The amount of inbreeding in a horse is known as the "coefficient of inbreeding". Keeping this below 12.5% or two crosses in three generations is best. Some recommend less, say 6.25%. But remember, there are no rules to this. Numerous calculators are available by Googling, so you do not need to work it out yourself.

Horse breeders can never in their lifetime inbreed to the extent that bird, mouse, or dog breeders can due to the generation gap, producing only one foal a year, and needing to get the desired sex to continue. Better to buy an inbred line and then continue inbreeding later when you have sufficient knowledge and experience.

There will always be recessive genes lurking in the background. How many of these are undesirable is unknown until they appear in homozygous form in the foal. So there will always be a threshold over which you get too many negatives. The trouble is that threshold is different for each population (or breed) within a species and different for every line within a population (breed). And it depends on how many recessives you (unknowingly) start with.

Mendelian inheritance shows 4 possible combinations of genes from each parent, when both parents carry a recessive gene. But two of the combinations are identical, so the ratio is 1:2:1 of probabilities. Three of these combinations will look like the parents, because they all contain the dominant gene, and only one will look like the recessive factor. Recessives can only occur when both parents pass on the gene.

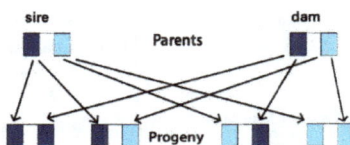

Our stallion Rannock was a half-brother x half-sister mating, and his dual grandsire Panzer's own dam was herself inbred to 1899 Sydney Cup winning mare called Diffidence (TB). We needed to check if Rannock had any undesirable genes. The only way we could do this was to inbreed with some of his progeny, which have to date been clear of negative issues. This is only possible because the breeder of his ancestors applied rigorous breeding principles at the time. Some of our horses, now have three and four crosses of Rannock in them, with no apparent effect. We endeavour to keep the percentage of Rannock blood as high as possible in our inbred pedigrees.

Our current stallion, Chalani Sunstream HSH, is the result of reciprocal back crossing, an inbred sire over an outcrossed mare. However, when one looks at his extended pedigree, he has the famous winner of the 1890 Melbourne Cup, Carbine in the background some 41 times. Additionally, Sunstream's dam is sired by a Rannock son. This means that his foals out of our Rannock-line mares, will be inbred to, or *mirroring* Rannock.

Chalani Sunstream has 41 crosses to Carbine (winner 2 Sydney Cups, and 1890 Melbourne Cup).

Inbreeding is a fantastic way to set type. You can concentrate the great qualities of a particular horse or line through inbreeding. But caution and review of results needs to happen with any programme, inbreeding or not!

With the early breeders, high levels of inbreeding reflected tradition and many generations of trial and error to get it right. They discarded those which didn't measure up. The King Ranch Quarter horses are a well-documented example, representing a policy of inbreeding true to type and for a purpose. There are many examples. Here is the campdraft horse, Lawlors Beaudana HSH. In addition, Dernancourt Manumber Lass carries Dernancourt Ben Abbey as her sire, which is Odet's dam's sire, while Commandant is Elliotts Creek Cadet's sire, and Odet's granddam's sire. See if you can work that one out.

Lawlors Beaudana

Foaled 2016
Blk Stallion

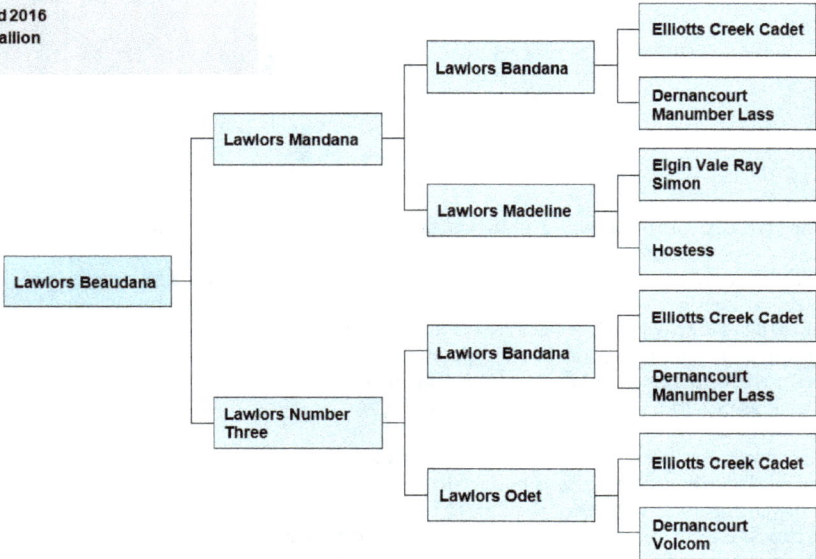

		Lawlors Bandana	Elliotts Creek Cadet
	Lawlors Mandana		Dernancourt Manumber Lass
		Lawlors Madeline	Elgin Vale Ray Simon
Lawlors Beaudana			Hostess
		Lawlors Bandana	Elliotts Creek Cadet
	Lawlors Number Three		Dernancourt Manumber Lass
		Lawlors Odet	Elliotts Creek Cadet
			Dernancourt Volcom

I believe a degree of inbreeding is less important than the individuals it creates. If genetic disease, deformity, or other frailties occur, outcrossing or complete abandonment of the line usually occurs. The pro is that if you're lucky and manage your breeding well, it's often possible to purge the line of some deleterious recessives and produce a line with "concentrated greatness," but without the inbreeding side effect.

Purging can occur because, in each mating, you lose 50% of the DNA from each parent. If that 50% held lots of damaging recessives, they are now gone forever, and that resulting foal can never pass them on. Over generations, with careful selection and culling, it's possible to "have your cake and eat it too." Geld heavily. Culling, for the breeder, merely means removing from the breeding herd. That is one reason to use sound and performed mares. They can be repurposed as riding horses if they prove unsuitable or end their breeding life. It is not always possible, but that is the goal.

> "The biggest issue we have now is good horsemen are a dying breed. The practising of culling hardly exists, and we are breeding horses that need to be put out to pasture." – Bronwyn Margetts.

When you find something that works, repeat it. You've already spent time and research making decisions that have worked for you. Unless you can develop a better option, why reinvent the wheel? It is easy to become obsessed with experimentation rather than being satisfied with the tried and true. On the other hand, some breeders just breed the same every year, even though the progeny are average at best. You know what they say about insanity!

It is important that you know how to look at and study pedigrees. Not only does it teach you about the horse, why some are successful, when others aren't, but it teaches you about how the good breeders *think*! Now, see if you can follow the pedigrees in the next pages.

Chalani Tempo ASH, stallion owned by Janita Edwards, here ridden by David Murphy 2021, a combination of the two foundation mares at Chalani. Chalani Cat Ballou is also the dam of Paper Tiger. Photo Jo Thieme.

Chalani Tempo

Foaled 2012
Blk stallion 15.2h

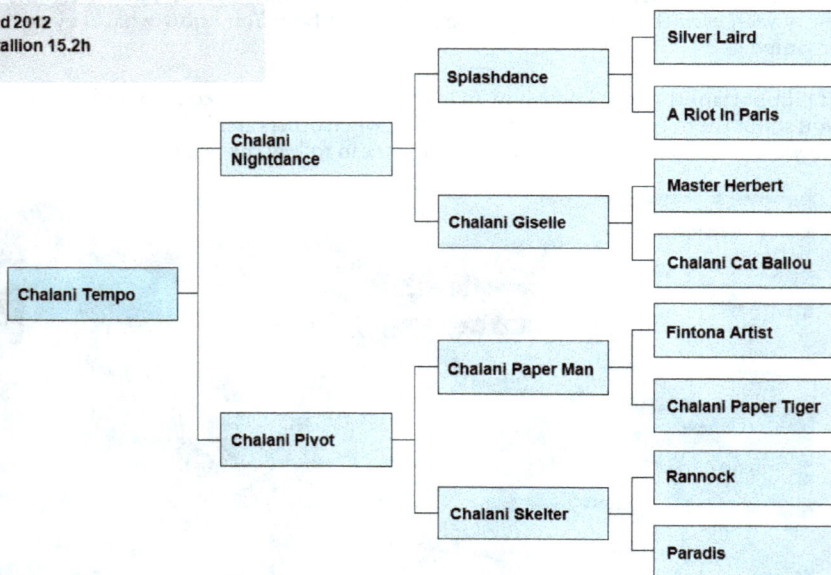

Chalani Tempo	**Chalani Nightdance**	**Splashdance**	**Silver Laird**
			A Riot In Paris
		Chalani Giselle	**Master Herbert**
			Chalani Cat Ballou
	Chalani Pivot	**Chalani Paper Man**	**Fintona Artist**
			Chalani Paper Tiger
		Chalani Skelter	**Rannock**
			Paradis

Rannock

Foaled 1967
Ch stallion 15.2h

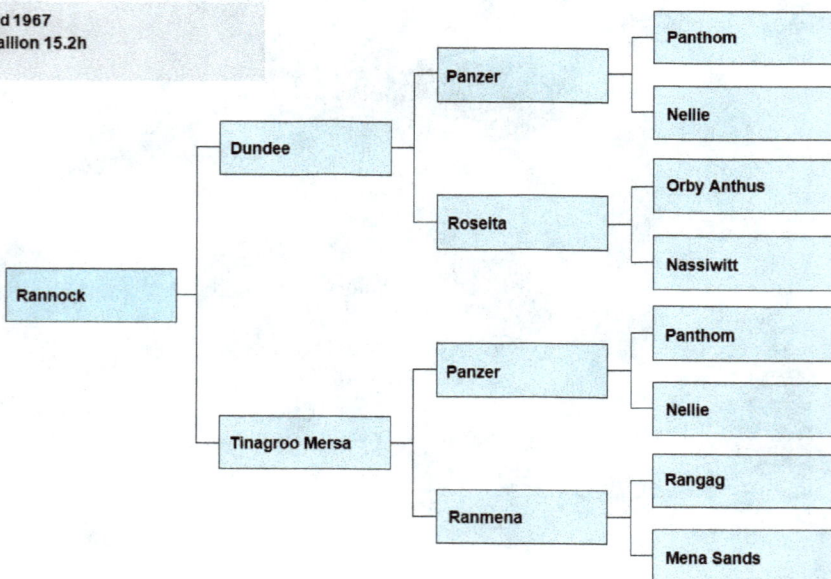

Rannock	**Dundee**	**Panzer**	**Panthom**
			Nellie
		Roselta	**Orby Anthus**
			Nassiwitt
	Tinagroo Mersa	**Panzer**	**Panthom**
			Nellie
		Ranmena	**Rangag**
			Mena Sands

In addition, Nellie, (dam of Panzer) is inbred to the 1899 Sydney Cup winner Diffidence, through Diffidences' son Kangon, and the tail female line of Gooroolai.

Chalani Mystic

**Foaled 1990
Gr stallion 15.2h**

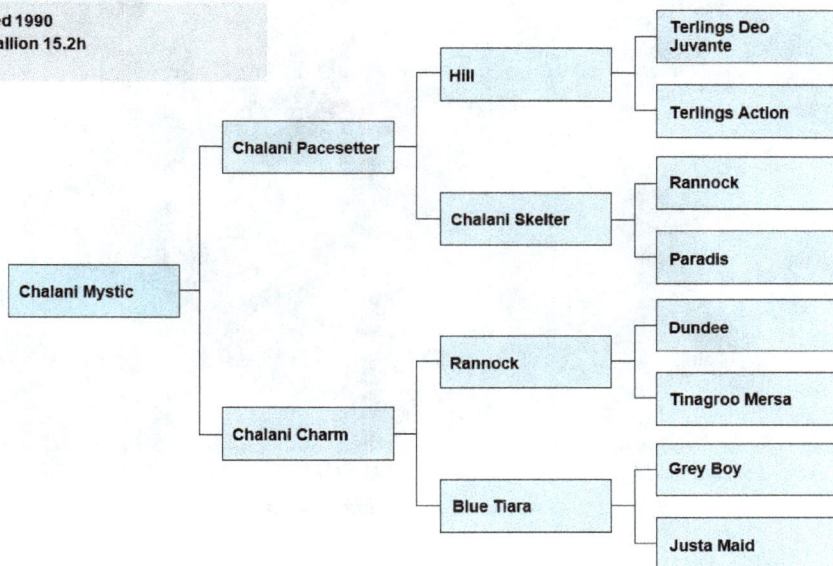

Chalani Mystic	Chalani Pacesetter	Hill	Terlings Deo Juvante
			Terlings Action
		Chalani Skelter	Rannock
			Paradis
	Chalani Charm	Rannock	Dundee
			Tinagroo Mersa
		Blue Tiara	Grey Boy
			Justa Maid

Chalani Galaxy

**Foaled 2021
Bay colt 15.2h**

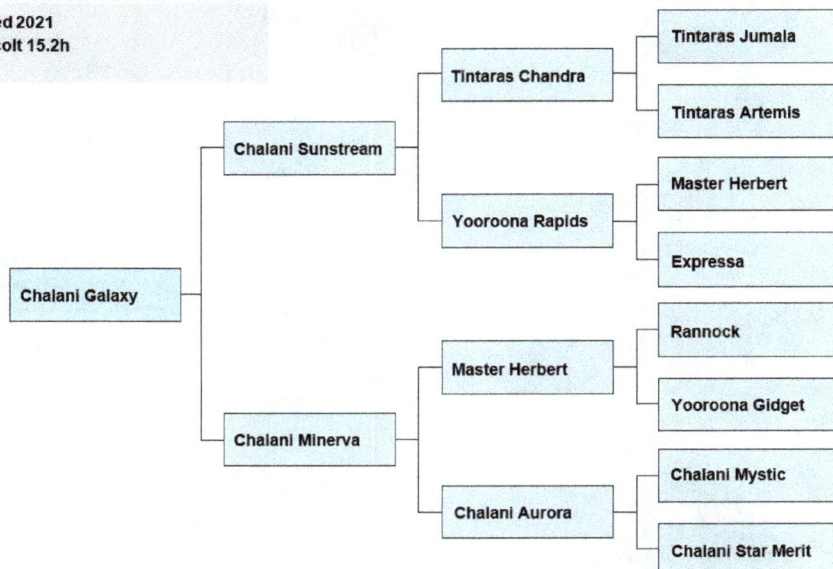

Chalani Galaxy	Chalani Sunstream	Tintaras Chandra	Tintaras Jumala
			Tintaras Artemis
		Yooroona Rapids	Master Herbert
			Expressa
	Chalani Minerva	Master Herbert	Rannock
			Yooroona Gidget
		Chalani Aurora	Chalani Mystic
			Chalani Star Merit

Hvirfs London HSH, with Lance Anderson, owner Ike Murray. Photo Justine Wake. A good gelding promotes the breed. And just look at that stickwork!

Hvirfs London

Foaled 2012
Grey Gelding

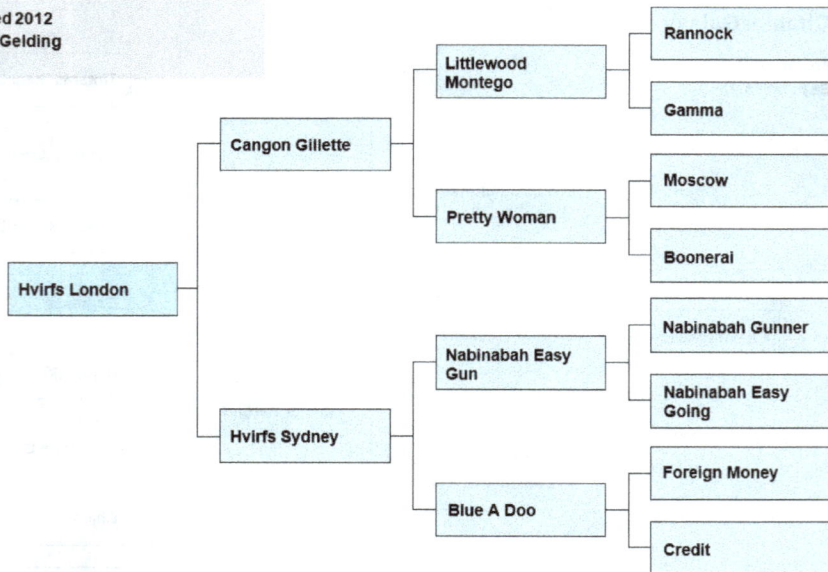

Hvirfs London	Cangon Gillette	Littlewood Montego	Rannock
			Gamma
		Pretty Woman	Moscow
			Boonerai
	Hvirfs Sydney	Nabinabah Easy Gun	Nabinabah Gunner
			Nabinabah Easy Going
		Blue A Doo	Foreign Money
			Credit

In addition, Boonerai is inbred to Sunny Hour and Credit is inbred to Bobbie Bruce, enhancing the value of these bottom-line mares.

Chapter 18

Inheritance

Genetic disorders

Genetic Diseases

Nearly every breed carries known genetic disorders, some extremely damaging for the breed or the individual. Some are "dominant" traits brought about by mutation in a recent ancestor or "recessive" traits likely in a more distant ancestor. Dominant traits must be inherited from one of the parents *in a single unbroken chain from the original source.* Recessive traits are inherited in double dose, *one from each parent.*

Six types of pairings and the possible ratios in the progeny.

More and more genetic defects are being discovered in horses, many similar to genetic defects in humans. A wise breeder will understand the inheritance of the traits within his own breed, and endeavour to educate clients and other breeders if asked.

The homozygous parents are shown in lines 1, 3, & 6. The results in 1, 2 & 3 will always be the same as the dominant parent, but they will breed differently. If breeding for type, or inbreeding, the result will be increasing homozygosity, in the direction of dominance, line 1, or recessives, line 6. In both cases, genes have been lost from the gene pool and can no longer appear until bred to an outcross to re-introduce it. This means once you have removed a genetic disease by testing, progeny are clear of it too.

I would recommend novice breeders steer clear of breeding stock tested to have genetic disorders. I don't hold the view that horses which are carriers have some special attribute that allows them to perform better than those which don't. Some traits are evident at birth, such as dwarfism. Others may appear later and be controlled to some extent by diet, such as HYPP (hyperkalemic periodic paralysis), which is characterised by muscle tremors, weakness and collapse.

Others have multiple symptoms which make them very difficult to diagnose without genetic testing, such as PSSM (polysaccharide storage myopathy), a muscle wastage disease of which there are two types. The second now has four genetic variants identified. These tests are important for diagnosis, so correct management can be formulated.

Then there are diseases for which there are no genetic tests, but are known to have high heritability levels, such as EMS (equine metabolic syndrome). I would question the use of such horses in a breeding programme, as the heritability is as high as 80%.

Breed panel tests are sets of tests relevant to a particular breed. Contact your breed association to determine what panel testing is mandatory for your breed. This is one reason to breed from and sell only registered stock. Today, there is no excuse for breeder ignorance of genetic testing. All breeders should have their breeding stock tested for the disorders within their breed unless the parents have already been tested clear.

Breeders should be prepared to offer the results of their tests. If they do not, avoid those breeders as unprofessional. Clients will need education about its importance and will want quality assurance.

With accurate genetic testing, disorders can be removed from the breed over time. But it takes effort and the work of responsible breeders, more so in the rarer breeds. There are so many good horses to choose from. There is no need to breed from a horse with a known genetic disorder when one of a similar line can be found without it. Tests are simple to conduct and inexpensive. Once you have sire and dam tested clear, their progeny no longer need to be tested.

For genetic diseases and panel testing information, go to the websites:

Practical Genetics (AUS) https://practicalhorsegenetics.com.au/index.php?test=qh5

Or the University of California, Davis Campus (USA): https://vgl.ucdavis.edu/tests?field_species_target_id=266

Breed Differences

There are considerable genetic differences between breeds, for example some breeds, like Lipizzaners and Minis are known to have shorter gestation periods than the norm. Minis are more prone to colic and foaling issues. Friesians are sensitive to anaesthesia. Some Arabians are one rib shorter than other breeds. Standardbreds, Morgans and Quarter Horses make the best blood/plasma donors as they are predominantly a universal blood type. Gaiting breeds and pacers can do so because of a single gene which determines it.

The fast twitch muscles of the Quarter Horse are suited to sprinting, and the slow twitch muscles to staying, while the "athletic heart" is linked to endurance. Cardiac measurements, specifically left ventricle measurements, correlate with performance and show strong heritability levels. Heredity influences fitness and exercise thresholds for aerobic endurance and variation has been found in response to training between breeds.

How does a novice breeder (or anyone) find out these little snippets about their breed? By doing their own research and asking other breeders. Anecdotal evidence usually precedes scientific evidence.

Some differences are the result of type (including colour standards), which have been selectively bred for generations to be "of that type" both for the breed's purpose and for its differentiation from other breeds. The type should be homozygous, that is, it will breed true to type once a type has been set. It is type for purpose (discipline) which sets type, not a registration. Sometimes different types which depart from the standard or purposeful use, may be found within a breed, but the original type should be safeguarded by breeders. If you want to depart from type, maybe you should be looking at another breed. When type disappears, it is almost impossible to get it back, even with several generations of crossing back to type, and then it may be too late, because the type may have been lost.

It is up to you as a breeder to know your own breed thoroughly so that you can breed accordingly, be a custodian for the breed, and provide relevant information to prospective clients.

Most breed traits show as a continuum. So at what point does normal variation become an abnormality? It has been shown with Russian studies on foxes, that breeding only for less "wildness," or aggressive nature over many generations, without selection for any other trait, changes the foxes' appearance to more dog-like, and they develop white markings similar to a border collie. This shows that traits are linked to other traits when we select for one thing. That is understandable as there are only 32 pairs of chromosomes in the horse, overall containing some 20,000 genes (or instructions), so when a chromosome is passed from one parent to offspring, many genes are passed on together by that chromosome. This is called "linkage."

"A little bit of a trait creates an advantage. Too much of a trait creates a disadvantage. DNA is the genetics which formulates the software; junk DNA (which was formerly believed to have no value) is now believed to be the operating system." – Temple Grandin.

So as we select for one trait only, we are doubling up on many other linked genes which we are unknowingly selecting for as well. Some may be faults, either mild or severe, such is the continuum, to a point where it can be impossible to return to "normal."

For example, mobility and flexibility is desired in the dressage horse, but so much emphasis on movement can mean breeding from spectacular, but hyper-mobile horses, which struggle to keep their own body stable. Some are even missing the lower portions of the lamellar nuchal ligament (currently under investigation in Germany). It is possible to breed badly conformed horses with the hyper-mobile gene, which are more "spectacular" than horse of well balanced, correct conformation and movement.

"We have domesticated the weaker animals as they were easier to subdue but in doing so we have created genetic deficits that are reaching catastrophic proportions." – Sarah Williamson

On the other hand, some genes are quite useful for either diagnosis, or providing a *tendency* in a certain direction. All racehorses have an optimal distance, based on their genetics, the so-called "speed gene". Though they might win over other distances, depending on preparation and competition, its genetic makeup foretells what its likely best distance would be. There is a "speed gene" test, developed in Ireland, which examines the myostatin gene, a gene responsible for the regulation of muscle development. Horses C:C had an optimum race distance of 1000-1600m (or 5-8 furlongs), horses C:T for middle distances 1600-2400m (or 8- 12 furlongs) , and T:T horses, the greatest stamina, over 2000m (greater than 10 furlongs). Of course, these are linked to other metabolic requirements and physiological responses which may be measured in fitness training, and are influenced by multiple modifying genes, yet to be determined.

Prove your horses before breeding from them, polo ponies here doing "sets." Photo courtesy K&R Polo.

The death spiral for any breed is breeding for single characteristics, such as gait, colour, size, speed, heavy muscling. With emphasis on heavy muscling in the Quarter Horse, we see animals at yearling age which look more like three-year olds. And by six-year-old, they have been retired to the breeding band.

How to ruin a breed

"First step: Breed the versatility out of them by sectioning different types to different classes in a show. Halter types look like a completely different breed than your ranch horses? Bingo!

"Second step: Breed the usability out of the ones meant to represent the breed. If your horse can go into a ring and make the breed look like it will absolutely NOT function past 8-9 years old, and you get the "vet fees are going to be high with this one" feeling, then you are on the right track! Remember, you just need pencil necks and tiny hooves.

"Third step: Brainwash the folks into thinking this is what they definitely must have in order to be a part of the "big boys" club. Charge them thousands for your dysfunctional weanlings that will need lip chains. Pay the judges to blue ribbon your train wrecks, so the usable horses don't place and "aren't wanted here." Those usable horses are only good enough for the trail riders now. Slap a catchy name on the HYPP positive babies for bonus points, like Built Lika Table, Jack Hammer, or Lame By Four.

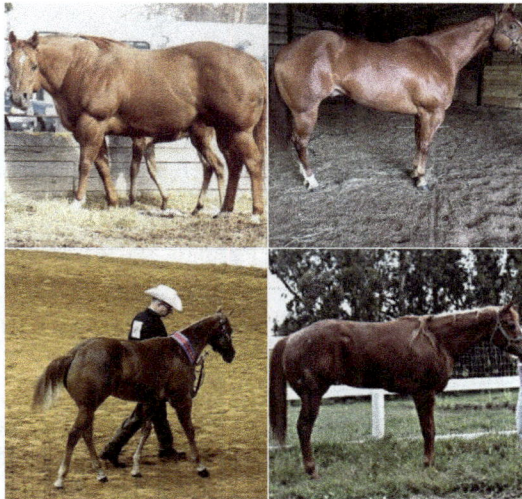

"Fourth step: Breed lots of mares so that you have lots to sell and then dump all those who didn't inherit the most godawful build you can imagine at the auction houses. Wean them at two months and start pumping your babies full of fillers and weight builders. It's okay that their tiny hooves won't be able to support the bulk you put on. You want your weanlings to look like two-year-olds on steroids.

"Fifth step: Call everybody a big poo head for not agreeing with you that the nightmares you created are the best thing since sliced bread. Everyone should want a horse that looks like a two-year-old. Scream, "this is how they are supposed to look." Go tell those with functioning senior horses that they "don't know what they are talking about."

"Sixth step: Sit on your throne, surrounded by delusional followers, and laugh in a low deep voice. You have successfully taken a functioning versatile animal and turned it into a walking vet bill that will need to be euthanized before you purchase a bag of senior feed.

"I'm not sorry if this offends anyone. I will absolutely stand up for the horse 100%. It's not fair to keep producing animals that will have a hard life because they aren't bred to last. If you are breeding conformationally sound, temperamentally sane, healthy horses that are true to their breed standard, then this does not apply to you." – Amanda Gray.

Colour inheritance

This is a subject close to my heart as I was an early campaigner for horse people to learn about colour, and wrote *Horse Colour Explained* back in 1999. Although Gregor Mendel theorised the existence of genes (or hereditary units) back in the 19th century, the chemical nature of genes, or DNA, wasn't discovered until Chargraff in 1950. I well remember the announcement that the double helix structure of DNA had been found by Watson and Crick back in 1971! Then in 2006, the horse genome was sequenced, mapping 2.7 billion DNA base pairs. Even this century, new colours have been described, whilst labelling of colours has been redefined and standardised, though local terms are still in use. So many advancements in understandings and tests have come in the last 20 years it is almost impossible to keep up, but suffice to say, master breeders of the past didn't need an understanding of genetics, or colour inheritance, to breed good horses of colour. What they didn't understand was the inherited anomalies which are now fairly well understood.

Grey is linked with melanoma, though breeding from lines which don't express it until very late, seems to reduce the risk of expression at an early age. Homozygous leopard horses, LpLp are linked with night blindness. White horses are a relatively rare sponta-neous mutation in the Thoroughbred of which there are several variations. Other white foals may be associated with the sabino gene. Roan has also been known to (very rarely) spontaneously mutate. Deafness can occur in Splashed white horses. Double silver (taffy) horses, ZZ, and chestnuts with double silver, may have MCOA, a congenital eye anomaly, which may impair vision.

Overo (or what some call Frame) carries with it a linkage to lethal white. Overo is dominant but the lethal factor is a linked recessive. That means all overo horses are heterozygous, or carriers. It is termed lethal because the white foals are born with severe bowel and other abnormalities which cause agonizing death within a few days, if not euthanised. This means breeders of this pattern must steer clear of breeding two overo horses together (or apparent solids tested to be carriers). The risk is 25% of producing a lethal white foal – Line 4 on the previous chart. (Lines 5 and 6 cannot occur because no adult can be double recessive).

Deformities

Sometimes you can be plain unlucky. Wry mouths, an extra or missing leg, hermaph-rodites, somatic mutations, missteps in the segregation of chromosomes, cleft palate, heart defects, disfigurements and deformities are noted in all species, and the horse is no exception though they are extremely rare. Perhaps this is because the horse is a prey animal, and many simply would not survive. The famous polo sire Norman Pentaquad had extra digits (hence the name – Google him for his story). However, science tells us that these are all developmental abnormalities which occur after conception. Such abnormalities are therefore curiosities and not heritable.

Chapter 19

Stallion profile

Ellerston Ruski

I profile this horse to demonstrate a stallion bred for the purpose, and selected by the Haydons for his full pedigree of high-grade polo ponies, his tail female line, and superior genetics. He is an outstanding example of good choices made by his breeder, Ellerston (Packer family) and owner the Haydon family, following the breeding principles in this book. Note the exceptional neck, slope of shoulder, elbow room and placement of stifle directly below the point of the hip.

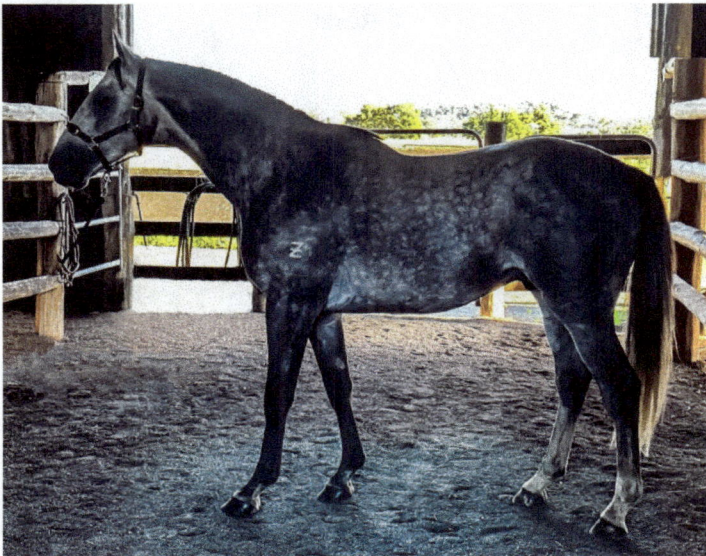

Ellerston Ruski HSH, cornerstone sire for Haydon Horse Stud.

Ruski's dam, grandam and great grandam were all played by top 10 goal players. He combines the two legendary mares La Luna, regarded as the greatest playing and producing mare in Argentina, and similarly for Pinky in Australia. It is not often you get a stallion with so many brilliant mares in his pedigree. This was only made possible by Kerry Packer's pursuit of excellence to acquire the best playing horses in the world. It is also not often you get a pedigree like this that is backed up with such great conformation, ability, action and temperament, that is passed onto his progeny. He also has the bonus of having an outstanding full sister, Ellerston Skyy. A rare stallion that ticks all those boxes.

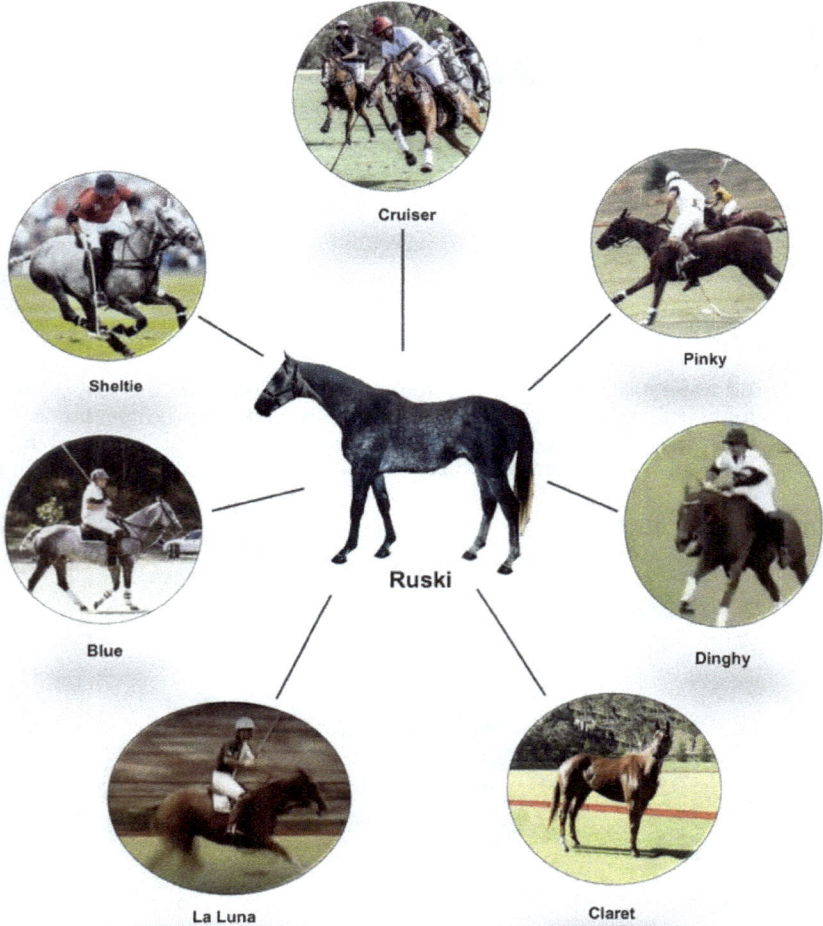

Cruiser

Pinky

Sheltie

Dinghy

Blue

Ruski

La Luna

Claret

Cruiser – Ruski's dam, rated as one of the best Ellerston mares, doing everything so easily at pace. In 2018 she won BPP on debut in UK and then the Argentine Open played by world's number one player Adolfo Cambiaso.

Sheltie – Ellerston Cavalier's full sister, a dominant polo mare played by 10 goaler Facundo Pieres for 16 years winning awards in both the English and US high goal.

Touch of Pink (Pinky) – A Hall of Fame mare. She was not only a top playing mare but also an incredible producing mare with an international legacy.

Dinghy – Ruski's granddam rated by Facundo Pieres as one of the best, dominating in Australia, England and Argentina for 12 years. Renowned for her extreme pace and incredible stop.

Claret – half-sister to Dinghy (by Littlewood Montego, by Rannock). Rated as Ellerston's best ever, she was an extraordinary polo mare who won the Argentine Triple Crown, was BPP in the English Gold Cup final and Ellerston final. Her progeny and many sire sons are used around the world.

Blue – became famous when played by 10 goaler Horacio Heguy and later by Kerry Packer. By the renowned polocrosse stallion Elmswood Punchline, sire of four times National's Champion, Leeway. Her dam is closely related to the dam of Narrangullen Anakin, 8 times top ten at the Man From Snowy River Stockman's Challenge.

La Luna- Legendary Argentine Hall of Fame mare and dam of Ellerston Solar, she has had such a big impact on world polo bloodlines.

ELLERSTON RUSKI

Foaled 2014
Grey Stallion 15.2

			Riverman
		Norman Pentaquad	
	Ellerston Cavalier		Lady Rebecca
			Elmwood Punchline
		Blue	
Ellerston Ruski			Cincinatti Aziza
			Optimum
		Ellerston Solar	
	Ellerston Cruiser		La Luna
			Longboat
		Ellerston Dinghy	
			A Touch of Pink

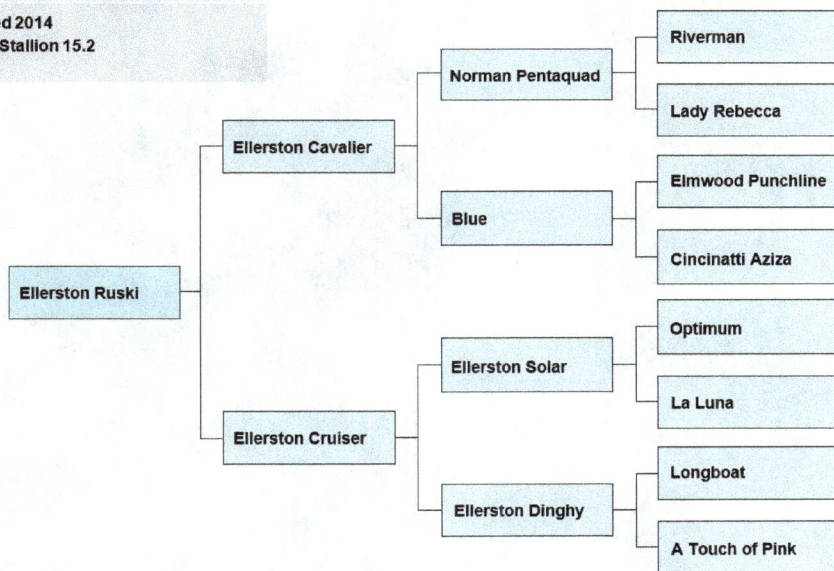

For Further information on the training and breeding programme of the Haydon Horse Stud, please go to their website. It is a mine of information; a prime example of a good website and how one can promote one's stud. It is one of the most successful studs in Australia.

Haydon Horse Stud: https://www.haydonhorsestud.com.au/

If you'd like a fillable copy of the design template for these pedigrees, please email me with TEMPLATE in the heading.

Narrangullen Anakin HSH stallion, 16.1h, at Man from Snowy River Challenge, riding with no bridle or neckstrap. Owner/rider Michael Green. They have placed Top Ten in this toughest of Australian Challenges no less than 8 times. This photo taken when they came 2nd in 2013. Photo Sarah Martin.

Anakin is by Witchetty by Rannock, (grandsire of Claret) and his dam Narrangullen Tikka has some rare common lineage with Ellerston Ruski, through Ruski's great grandam, Cincinatti Aziza.

Chapter 20

Choosing the stallion

Selecting the right stallion for your mare

Breed the best to the best and hope for the best.

In its simplistic form, the sire and dam must be the best you can find and the best fit. The stallion should complement the mare. You should not breed her to a stallion with the same faults. Evaluate that on all levels, from temperament to conformation, to pedigree. Both should be as close as possible to your ideal type. The right stallion for your mare comes down to how complementary they are. You want to maximise the good qualities of both and reduce any of the lesser qualities. This makes sense, so how do you find *the* one? Start looking for a potential sire well before the stud season. What is the purpose of the mating? To keep for yourself, or a commercial proposition?

Compare the service fee against the value of the progeny. For example, if the highest price is around $40k, but the average is $10k, it is likely not worth paying $3k for a service fee, unless your mare is of the calibre of the dam of the $40k horse. If the averages are very consistent, say, $30k, you have less risk, and more likely a profitable venture. Also double-check the age and training of the highest priced horses, to see if you can match that.

If the stallion is in his first few seasons, with no progeny, or none which have performed, he must be evaluated more as an individual and on pedigree. All the more important that he himself has performed! Occasionally you might choose one of exceptional breeding before he has been taken out, in the hope that his progeny will be sought after later.

Firstly find horses of the type and discipline you are looking for in regards to conformation and movement. Look at their pedigrees and check the extended pedigree. Make sure you know most of the names in it and what they have done. Will the foal be inbred, or outcrossed? Are the bloodlines complementary? Make sure you look at the stud's website. Check if progeny are registered. If other breeders are not using the horse and registering them, find out why.

Using the information in the previous chapters, make a list of suitable horses, together with your comments and the stud fee. Only list those you think are a suitable match, and reasons. What would the match bring to your stud? If they are out of your price range, still list them, because you never know, there might be a charity fundraiser in which the stallion is offered at a discount.

Don't just look at advertised horses; start looking at the sires of winning horses, and ask around your local area. Some very good stallions stand at moderate fees, but go under the radar, simply because their owners do not advertise extensively. Go to some local or state events to see what is out there. If you see horses you like, find out their sire by talking to owners. Owners are usually quick to tell you about quirks and give recommendations.

Create a short list of at least ten horses. This is when you start looking at the facts in greater detail by contacting the stallion owner, asking any questions you might have and seeking contracts. Will they be standing at the owner's property or at a breeding centre? Ask what they know about the sire and dam. Ask to visit or view video. See foals or progeny and ask what they have done.

This will also give you an idea of the communication from the stud owner. He will be less busy outside the stud season so should be able to reply to you within a suitable time. Think of your questions in advance, so that you are not bombarding him with many individual questions which might become annoying.

Be honest with the qualities of your mare, so that you can appraise if the stallion is suitable, particularly as to idiosyncrasies, and anything which bothers you about her. You want to know if the stallion you choose has a good chance of overcoming these. You also want to know if someone with your experience (or lack of) can manage the resulting foal.

Never breed a large stallion over a small mare, especially a maiden mare. Apart from the distress to the mare if she foals an extra large foal, complications are costly, and may result in death of the foal, or both animals. This is a welfare issue. So how big is too big? The difference between stallion and mare should never exceed 2 hands. This should be the absolute limit, and preferably *far* less.

Various reasons have been found why people chose not to use a stallion or buy his progeny. Some of the reasons have nothing to do with the animals themselves! You can decide for yourself if the reasons seem valid. Sadly, if a stallion has an off day at the wrong moment, it can stick in the memories of people who may only ever see it at that moment!

Reasons against the stallion:

- Did not know the stallion existed.

- Thought the stallion was infertile.

- Sent the mare to the stallion once and was unsuccessful, so tried a different stallion.

- Too far away.

- Too closely related to the mare.

- The stallion was the wrong colour for them.

- Viewed the stallion but did not like his temperament/conformation/movement/ progeny on the ground.

- Did not think he would suit the particular mare.

- Did not like a relative owned by a friend.

- Saw the stallion (or his progeny) at a show or event and did not like his behaviour.

- The mare owner did not like the stallion owner.

- Did not like the stud's setup or how the stud owner cared for their own stock.

- The stud fees or agistment were too high.

- The breeding contract was too complicated.

- History - failed licensing, failed x-rays, not being licensed or licensed differently because the first lot didn't pass.

- Genetic conditions.

Secretheart Silver Echo HSH taffy stallion. Photo by Sharleen Flannagan.

Here are some reasons why people *do* choose to breed to a stallion:

- Knew of the stallion and had seen him competing and loved him.

- Researched the lines which would complement their mare and hopefully introduce some improvement.

- Went to see him in the flesh and liked him.

- Had seen multiple progeny and liked them.

- The stallion owner was known to be experienced with breeding, have good husbandry techniques, and look after their visiting mares well.

- Had good fences for visiting mares who were in an unfamiliar environment.

- Recommendation by a more experienced breeder

- Live cover was chosen as the stallion was accessible, or the mare had been tricky to get into foal frozen or chilled previously.

- The owners had DNA tested the stallion and mare, and no known genetic disorders had to be considered.

- He was "trendy" at the time or believed the marketing "hype."

- Were impressed by the advertising or the stallion's record.

- A live foal guarantee was written into the contract.

- He had flashy movement.

- He was a particular colour or pattern.

"Temperament is something that is different for various owners. Tulira Colman (Connemara) was thought to have an excellent temperament by Lady Hemphill, but when he moved to the UK, his owners (used to handling a stallion who was a total teddy bear) could not handle him, and he got away at a big show in the UK and was a bit of a disgrace! M saw him and loved him immediately (and she was a real go-getter like Lady Hemphill) and thought he was the most brilliant pony. And he went on to being ridden by her children too.

"She later purchased Connaught of Millfields because she thought he would be similar and a real performance horse and pony producer, but when she owned him, she thought he was a total slug and a next to useless stock horse!!! But he probably had an ideal temperament for less gung-ho riders.

"No one can be expected to get along with everyone who shares their interest, and it would be great if people could set aside personality and just do the business, but it does make a difference in choices people make. Sometimes there may be a few factors which sway the choice in a certain direction." – Tearna Golston.

Choosing a stallion from a photograph is difficult. There would hardly be a stallion advert in some disciplines which has not been doctored in some way, sometimes to the point of misrepresentation. Some are easy to pick, for example, the photo has been tilted. Ask to see the original photos, or video of the horse, or better still see him in the flesh, where you can ask to see him move, and ridden if possible. If he has been posed to the extreme, ask to see some more natural photos.

There are also many stallions retired because of soundness issues so it is important you seek second opinions from respected parties, such as the trainer or owners of progeny, as to why he might have been retired (or retired early).

Before proceeding with the stallion of your choice, do check that all its stallion registrations, membership fees, mandatory testing, and breed requirements are up to date. The owner should give you all this information but double-check against the breed website. If not done, these are red flags.

Don't believe all this will be done "once you have paid the service fee." Read the fine print. Check the contract for when you will get a Service Certificate. Technically, the stallion owner must give this to you upon service (or 30 days upon request), not wait till you have a live foal. It is not a *live* foal certificate but proof of *service*. However, the owner is not required to supply it if you have yet to pay your bill in full!

Check when payment is due. It may be at 14 days, or at 45 days, or when you take the mare home. With AI you will be asked to pay up front, so check what returns may be available, if the mare is not pregnant first up.

Chalani Tempo ASH stallion owned by Janita Edwards. Photo Kelsey Stafford.

Narrangullen Anakin HSH stallion, makes his way to winning the "Brumby Catch" - Man From Snowy River Challenge, 2013, with Michael Green. Photos Sarah Martin.

Chapter 21

Stud work

A good relationship with an excellent reproductive vet is vital to a successful breeding operation. He is being paid for skills, not just time. The trust needs to work both ways.

Unless you have a large farm where the stallion is with the mares at the beginning of the season, it is usual to have all empty (barren) mares checked by the vet upon the first cycle after winter to establish her breeding soundness, and if any treatment is required. Many mare owners do not recognise when their mare is cycling because she does not show readily or they have no suitable horse for her to show to.

If you are sending a mare to the stallion, try to record her dates, prior to sending her, so the stallion owner has a history and give him the results of any tests. Better still if you can provide a brief written history of previous breedings and results. We can print ours out from HorseRecords.

A normal mare will come into heat or be "on" every 21 days and stay on for five to seven days, when she will ovulate 24 hours before no longer accepting the stallion or going "off." Just coming out of winter, she may be considerably longer and may not even ovulate. This is known as the "spring heat."

The service or insemination is timed close to ovulation. Assuming the mare conceives, she can be manually tested with ultrasound from day 14-18. The reason for testing early is to detect twins or any other issues. One twin can then be squeezed, or the mare can be re-bred while still early in the season. One would then check from 25-30 days for twin loss and development of the embryo. If fluid builds up, it needs to be treated early or it will become extra difficult to treat.

If the mare has a foal at foot, she usually cycles seven days after foaling, but this is very much mare-dependent. If she retains her afterbirth, develops an infection, tears, or has any other foaling problem, it is wise not to breed her on foal heat. It is not unusual for her to have a thick chocolate-coloured discharge for a few days after foaling. This is not concerning if it has cleared by foal heat. Most breeders avoid breeding on foal heat unless the mare is foaling late.

If she needs to recycle sooner, it is possible to "short cycle" her with a progesterone injection. Some mares will not cycle at all after foaling or after foal heat. This is known as lactation anoestrus. These mares are known as "every second-year mares." This is inherited, and she will likely pass it down to her daughters and her sire sons may also pass it on.

The most common vet procedure on a stud is follicle testing to determine if the mare is developing a follicle of a suitable size and when she is likely to ovulate. Mares' follicles typically mature to a size between 40 and 50 mm before ovulation, somewhat smaller in ponies. Mares don't read the manual, so it is not unusual for them to do something different. It is an art, not a science. Sometimes she will even ovulate up to 24 hours after she stops showing to the stallion. She will readily show her disinterest. A mare should only be bred when she is "on" because that is the only time her cervix is open.

Skyview Stud's Skyview Champagne Grace HSH (mare) and foal Skyview Honour HSH. Photo Jenni Phillips.

Difficult mares:

I disagree with forcibly breeding a mare that appears to be difficult. If she has been teased in a gentlemanly way by the teaser, or the stallion she is to be bred to, she will happily accept him. Teasing takes care and time, especially with maidens. Some mares are quite obviously on but will not accept being rushed. In all my years, I have only come across two mares that required twitching to be served. These were fearful of a stallion from past traumatic experiences.

Another mare had been running with a stallion for five years and had never gotten in foal. We decided to hand-serve her to see what the problem was. It appeared she was very ticklish and couldn't stand the stallion's legs around her. We put a rug on her, and she was bred within five minutes. She got in foal.

Another mare would duck her hindquarters away from the stallion at the last minute. This happened time and time again. She wasn't dangerous, but she knew something was wrong, though we didn't until a vet examination. She had a very narrow pelvis, most likely from malnutrition in drought as a foal. Unfortunately, she wasn't a candidate to have a foal. It would have saved the owner a lot of time had he checked her out before sending her to stud.

Most mares will be checked again at 45 days for embryonic loss. We advise the mare owner to do this because, by this time, the mare has usually returned home.

When should you breed the mare?

This depends on personal preference, but the larger the stud and the more popular the stallion, the more likely he will be restricted to "the season." A horse's birthday for most breeds is from Aug 1st in the southern hemisphere and Jan 1st in the northern hemisphere. This means studs operating on a commercial basis like to start breeding mares from the 9th of September and the 9th of February, respectively. Anything earlier, and the mare could quite possibly foal too early. Commercial studs will usually close off the season at a set time, so make sure you check when this is. There is no point in sending a mare on her last cycle before the closing date. If she doesn't catch right away, and many won't because they become unsettled on arrival, you miss a whole year.

It is best to have the same place where you serve your mares. Use a serving halter/bridle, which is different from when the stallion is ridden. This is so he is habituated to one place and doesn't expect to serve anywhere else. It helps him to be a normal horse outside his breeding shed. Use a longer lead when handling a stallion or serving with him. Both mare and stallion need stable footing to avoid injury. A mound may be required for height differences between the mare and stallion. Stallions tend to stretch muscles serving, especially in the shoulders and pelvic area, so it is good to have a body-worker check him out before and during the season.

The easiest and most efficient method is having a small, railed yard next to the stallion's yard. The mare is led into the yard and allowed to sniff the stallion, and he can do the same over the rail. If you think she is on but reluctant, she can be left in the yard with some feed while you do other chores, all the while keeping an eye on her behaviour. She will usually approach him a little later of her own accord and then show properly. We hold the mare on a very long rope through the other side of the fence so her chest is against the fence while allowing the stallion to enter through his gate. He then free-serves her. If the stallion or mare is likely to be difficult in any way, we would use a second handler.

Some mares will not be happy with you teasing them while their foal is close to the stallion. If there is a foal at foot, it can be contained by a handler or in a yard next door.

Signs of readiness:

The mare relaxes her tail, spreads her legs in a relaxed fashion, winks her vulva, or urinates. She allows the stallion to check her all over, particularly her neck, shoulders, flank, and hind quarters. She will usually swing her hindquarters in towards him when she is ready for him to jump. Some will ignore the horse but stay in his "zone" while eating grass. The stallion will turn up his nose in Flehman's posture and show genuine interest. These are normal behaviours. The longer the mare has been on, the more likely she will show textbook behaviour. When she is coming on or going off, she may squeal or be restless around him and kick out.

Signs of the mare not ready to be bred:

She becomes tense, flattens her ears, squeals and tries to run away. She may wink her vulva and urinate, but this is a sign of fear or anger, not readiness. If she does this, she will flick her tail from side to side, somewhat like a cat. She will kick out behind or lash out with her front leg. The stallion decides after a relatively short time that he is disinterested.

If the stallion is a commercial horse, that is, has more than 30 mares per season, it is usual to have another horse, known as a "teaser" to check the mares. This may be a pony, which can be relied upon to sniff and show interest in each mare brought to him. If the teaser is a nice type, he is often offered to outside mares at a reduced fee, so that he keeps his interest.

Stallions which aren't allowed to tease their mares learn to jump on the mare the minute they see her. This is quite dangerous for horse and handler. Serving hobbles or kicking boots are a necessity under this regime because you will never know if a mare will resent this lack of courtesy until she shows you in no uncertain terms. We have found that supposed difficult mares are frightened or traumatized by past stallion experiences. This should never be allowed to happen.

The stallion is allowed to stay on top of the mare until he relaxes his penis. Separate the two immediately. If he should be inclined to grab the mare's neck in order to balance himself, you can use a muzzle or buy a neck collar for him to grab between his teeth. He should never be allowed to bite the mare.

We don't free-serve with a young stallion. He needs to become experienced first. A good stallion handler is a blessing. The young colt may show a lot of behaviours before getting the job done. Some even take the whole season to get it right. They may bite, kick out, get excited without becoming erect, or jump on the wrong end while becoming silly or disinterested. Retired horses that have not been allowed to think about mares during a performance career, or those only used to paddock mating, may not be willing to breed initially. It takes patience and a patient mare.

Some owners dislike serving with a 2-year-old colt and prefer to wait until he is settled as a ridden horse. We have found it makes no difference. At least if you use him over a few mares as a two-year-old, as a test breeding, you will find out if he is fertile and what he can produce before you go on to a costly campaign with him.

Under natural service, a stallion can usually cover two to three mares in a day for extended periods. Mature paddock stallions can be run with up to 30 mares over a season with good results if there is sufficient area to house them all.

A stallion paddock mating will naturally service mares at the beginning of their cycle. He may serve her as often as every hour. If he has others to serve, he will become excited at the next one to come on, often leaving the first. Sometimes you will not even see him serve a mare as she may only make advances during the night when the other mares are quiet.

Watch out for any mare monopolizing the stallion, who won't let others approach him. If he is mobbed by several mares in heat, he tends to only serve his "favourite" regularly. Anecdotal evidence suggests more colts are produced when the service is timed with ovulation. This is thought to be why paddock mating produces slightly more fillies than colts. I have heard of quite a few stallions that will not allow a new mare into the herd or will not breed them until they are socialized in the herd. What matters is to know your stallion.

Our routine is to breed the mare shortly after she comes on, or when she arrives at the stud, in case she goes off before you can breed her again. We then breed every second or third day. We have found most stallion semen will last four to five days in a healthy uterus, without re-breeding. This is important with those mares which go "off" before they ovulate, which is thought to be about 10% of mares. One can then elect to leave her in with him for paddock mating, if desired. Make sure you have this in the contract, if you allow paddock mating.

Mares may become pregnant following insemination as early as 6 days prior to ovulation, and there are records of long sperm survival or delayed embryonic development, (up to 26 days) with DNA proof of the first stallion, (sire A) being the sire instead of the second (sire B). Industry recommendation is every second day, though if a mare is difficult to get in foal thus requiring medical treatment, it is better to only breed her the once, as recommended by your vet.

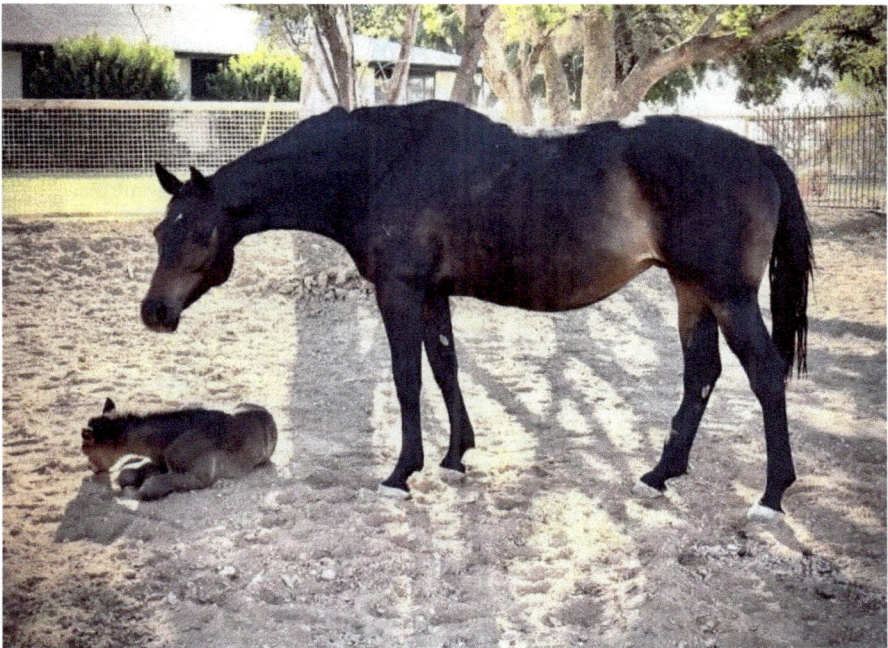

Chalani Maia, ASH. Photo by Kelsey Stafford.

Commercial studs rely on the vet to advise them when to breed, and that is the only time the mare is bred in her cycle. Unfortunately, only the racing industry provides statistics such as conception rates, number of starters, progeny, etc. So if sending a mare out to a stallion, you must enquire about conception rates. Natural service has the highest conception rates, but with improvements in AI techniques, the difference is no longer of concern. It depends on stallion fertility and vet experience rather than the method itself. Frozen semen is also closing the gap.

Stallions used for AI collection, on or off-farm, will require some training to a "dummy" or "phantom" at an experienced clinic. Ideally, before he has become experienced at natural service, otherwise it may be harder to train him. Some studs never do natural service with their stallions. If a dummy is not available, an in-season mare is used. Freshly collected semen is evaluated and then inseminated into mares. In a stallion of normal fertility, the semen from one ejaculate can be used in two to three mares simultaneously, if required. Or it may be frozen for future use.

We prefer to send/receive frozen semen, as chilled can be fraught with many difficulties, ranging from transport hold-ups, stallion-mare owner connection, or lack of skilled personnel. With frozen semen, it can just be taken out of the tank when it is needed. If the stallion is sold off or dies, you still have it.

Owning a stallion that you hand-serve means you don't get a day off. On Christmas day, if the mare needs breeding, you just do it. If the stallion is out competing, make sure that the stallion will be home when the mare needs breeding .

After all is done, the services and vet activity must be recorded. We prefer to do this on a programme like HorseRecords, designed as a user-friendly platform for small and large breeders. The larger you become, the more complex the paperwork and the greater the need for good records. Even if you paddock mate, check daily, record all mares in heat, services sighted, and certainly the entry and exit of the mare over the season.

The intricacies of breeding from a stallion may become too large and complex for the small stud, so perhaps send him to a breeding centre to stand the season. They will have the expertise and the staff. It may also make him more accessible to mare owners. This is a good option if you can find the right facility. Unfortunately, a facility can lose a good reputation quickly with management changes, so choose carefully. Some places have regular clients and good promotional opportunities, expanding the range of available mares.

Ironhorse Uptown Girl, Appaloosa, with Ironhorse Bo Jangles, foal. Photo Iron Horse stud.

Chapter 22

Care of the broodmare

Good management starts *before* conception. Ideally, the mare is on a rising plane of condition just after she has lost most of her winter coat. However, this is not an excuse for having the mare in poor condition over the winter. A mare in good condition all year round but not fat is good. Fat mares, particularly those that have been fat for a long time, are generally harder to get in foal. In addition, they may experience an overload of protein at later stages of gestation, which is not good for the developing foal and may be a significant cause of bent and crooked legs.

As the days grow longer, the mare responds by dropping her coat and coming into season. Blue light therapy may be useful early in the season, using a blue-light mask. The artificial light aims to improve fertility by stimulating follicular development, early breeding cycles, and coat condition. It provides the optimum level of timed blue light to a single eye, replicating the benefits of long summer days. It needs to commence eight weeks before the start of the season. The mask is quite expensive, so a better option for multiple mares may be to put them under lights in a stable or shelter leading up to the start of the season. You will need 16 hours of bright light daily, so setting the lights up with a timer is best. The light must extend to every nook and cranny of the yard or stable because the light must reach the eye.

Schedule a pre-breeding ultrasound exam (scan) when she first comes into season. This will include the state of her uterus and ovaries and show if any follicle is developing, or if cysts or other irregularities are present. If her vulva is too sloping or sunken due to age, conformation, or having had many foals, she may need a "breeding stitch" after service or a "caslicks" procedure upon confirmation of pregnancy. The upper part of the vulva is sutured closed to avoid an infection in the uterus.

After our mare Chalani Cat Ballou foaled her first, our vet suggested caslicking her, and once done, it needs to be repeated with each foaling. By the third time, the local anaesthesia was having limited affect, so I called a stop to it. We would take our chances, so as not to put the mare through it again. She ended up having 16 live foals.

A non-maiden mare should also have a negative culture by swabbing when she is in season. She should be cleared of infection before being bred, as infection can affect her long-term reproductive health. This should be done as early in the season as possible to allow time for treatment.

If the mare hasn't come into season as expected, consider getting her pregnancy tested. It may be necessary to give her progesterone to initiate cycling.

If the mare conceives, her pregnancy will last 11 months and 11 days. Therefore, 340 days, give or take two weeks on either side, is considered normal. Some breeds have shorter gestation averages. Multiple gestation calculators are available, but we use the free HorseRecords calculator. Any foal born before 310-320 days is unlikely to survive.

Once pregnant, the mare should go into a paddock where she can remain as calm as possible and be routinely checked and fed as required. We don't rug our broodmares unless they are aged and need extra help to maintain weight. Mares with foals at foot should be cared for in their own paddock to benefit both mare and foal. If you have Thoroughbred mares, they will likely need feeding extra on top of what your other mares require. Many mares lose their top lines while feeding a large foal but should never be allowed to become skinny. This indicates that they are not getting enough protein and bulk. Lucerne hay can provide both.

Chalani mares and foals

Can you ride a pregnant mare? Absolutely. If you have ridden her regularly, there is no reason why you cannot continue to do so until she becomes too heavy or sluggish, but I wouldn't jump her. What if she has a foal at foot? In some countries, this occurs regularly, especially with carriage or draft horses. Foals often accompany their dams in work. But you do have to feed accordingly and allow opportunity for the foal to suckle.

Ensure dental work and worming is current and have their feet in good order throughout. I never cease to be amazed at studs that will showcase their show stock but leave the broodmares out in the back paddock looking dreadful. The care of your broodmares is an indicator of the attention you pay to your horses as a whole, and I believe, the reason why we can sell most of our stock as foals at foot.

Foetal abortions, usually occur between the fifth and seventh month, but can occur at any time. If this happens it is unlikely the mare will experience any discomfort, and you will be none the wiser until she shows back in season at the end of the winter. It is possible the owner may notice development of the udder, some labour signs/colic, or discharge after birth.

As the pregnancy enters the last trimester, watch out for any unusual changes which could indicate a potential problem, such as oedema, running milk prematurely suggestive of premature foaling, discharges, herniation, etc. These require immediate attention from your vet. One of the most difficult conditions is placentitis, where the placenta becomes inflamed or infected. This condition commonly occurs in middle-aged mares that have produced several foals. It requires continual monitoring and treatment, or she will lose the foal she is carrying.

The occasional mare will show stallion-like behaviours when pregnant, particularly if she is paddocked with an in-season mare. However, this is quite normal and will disappear closer to foaling.

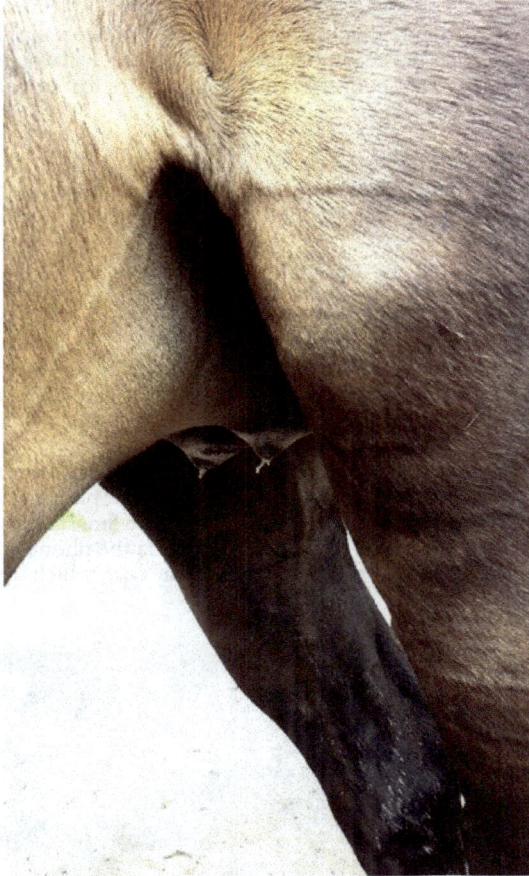

Classic waxing, approximately 12-24 hours before
foaling,

About four weeks before foaling, it is recommended to give tetanus vaccinations to allow immunity to pass to the newborn via her colostrum. It is time to unstitch her if she has been "*caslicked*." A good rule of thumb is when the udder begins to enlarge, or "bag up," she will foal three weeks later, though this is certainly not so with all mares. Seven days before foaling, you might see colic-like symptoms as the foal moves into the birth canal.

Unfortunately, mares who stream milk for several days before foaling may lose large amounts of colostrum, the vital first milk containing antibodies for the newborn. Now is the time to source some, rather than when you need it urgently.

Most mares will follow the same pattern each year. If a mare tends to foal early or late, she tends to most years. It is good to keep a record of gestation length for all your mares, so you can separate them when foaling appears imminent. We like to do this three to four weeks before foaling, usually into a paddock with other mares to foal around the same time and as close as possible to the house for observation. In the last month, she might become very uncomfortable and do much standing around. But it is better if you encourage natural movement, like keeping a distance between the feeder and the water trough.

Some breeders on large properties will allow the mare to foal naturally and unobserved, in with the stallion, but this is not recommended for more intensive operations. The mare doesn't need unnecessary distractions. Never let your mare foal in with a gelding or an aggressive mare.

You will notice her pelvic muscles slacken or soften, especially in the last 24-48 hours, preparing the pelvic ligaments for foaling. This is more extreme in older mares. If you feel along either side of her tail, you will begin to notice the changes before they are visible.

When foaling is imminent, the normal mare will have a tight udder, which may even look bluish, and peep out behind her back legs. She may have slight droplets of milk on her nipples in the last day or so. If it is sticky and solid, like candle wax, she has "waxed up." These are good indicators that the mare will foal in the next 24-48 hours. It is now time to put on the foaling alarm if you have one. Some breeders swear by foaling test kits, but we have never used them. It is good to have the paddock floodlit at night. Otherwise, accustom new mares to the torch!

The simplest alarms are those attached to the halter and placed on the mare up to 7 days before she is due to foal. The mare camera view is sent to your phone. There are several on the market which use radio waves to beep a base unit. The Smart Foal App is similar but incorporates the app, activity monitoring and alarms via the phone as well as base unit. When a clinic foals down, they use the "sew in the vulva" type, which trips the alarm when the vulva is broken. It is basically infallible.

Arabian broodmare Kalida Al Shaklan and foal, Karanji True Delight. Alarm attached to halter. Owned / photo by Karen Carr

You can now get excellent WiFi cameras off Amazon, to monitor foaling. These are perfect for mares foaling in a stable or yard close by.

There are many other monitors and systems which can range from in-expensive to very sophisticated, well beyond the scope of this book. The benefits are readily researched by reading reviews and contacting the supplier.

Foal App alarm system   https://www.foalapp.com/

Arabian mare Fairview Amira Shaklan owned by Jessie Rae Preece. Photo by Katherine Szalay Evans.

We prefer our mares to foal in the paddock, as most are more comfortable in the open. However, in bad weather, we may bring the mare and foal in. It is important to prepare an undercover area before needing it. A large foaling yard is ideal, with an open shed on one side containing new straw, shavings, or sand bedding. This must be as clean as possible so the mare and foal don't pick up an infection. A stable should be larger than a conventional stable, at least 4.25 m x 4.25 m (14 ft x 14 ft). In an emergency you can partition off your hayshed or carport!

As foaling draws closer, the mare will become most uncomfortable, pawing, sweating, and circling. She may get up and down and even roll to assist in positioning the foal. She may do this for some hours or be down and foaling without much sign. Once the waters break, foaling should occur almost immediately.

The broodmare will foal very quickly, and it is not always possible to be there when she foals, despite your best efforts. You may have her in a stable and sleep next door, and still miss her when you get a coffee! If that is the case, she nearly always foals without complications. It was 10 years before I saw my first foal being born.

Your broodmare is the most important asset on your stud. Treat her like royalty.

Top: Reysn Hope, Quarter Horse gelding, with Wayne Hinder. Ken Anderson Photography.
Below: Lil Rey Of Hope, QH mare 2012 NCHA Futurity Champion with Jason Leitch.
Wild Fillies Photography. Full siblings bred by Karen Thrun.

Chapter 23

Foaling and the first few days

Now comes the exciting part you have been waiting for, the foal's birth. Most mares will foal in a textbook fashion, and the foal will be up and drinking within two hours. Mares give birth quickly, which is nature's survival mechanism against predators. Compared to cattle, the ratio of foaling complications is much lower, but if this is the one time you have complications, you need to be very quick to determine if assistance is required. Cattle can go for hours with a difficult calving, probably with both cow and calf still fine. Not so with horses. If something doesn't seem right within 20 minutes, it probably isn't. Seek immediate help.

If you doubt your ability to handle complications, or your nearest vet is miles away, I suggest you book the mare into a reproductive clinic that provides a foaling-down service for a fee. If complications are anticipated, the mare can be safely induced with a low dose of oxytocin. This does not guarantee a live foal but gives you some peace of mind that the best help is available.

Mares who foal early don't tend to have as many complications as mares who foal late. This is because the foal is generally a little smaller, thus making birth easier. But mares do not follow the rules, and anything is possible. I've learned that every breeding season is a new challenge and a constant learning curve.

If you are attending the birth yourself, have hand cleanser, gloves, a clean towel, cotton wool, and iodine on hand. Clean hands are essential.

So what is "normal?" The beginning sign is the presence of the foal "bag" which contains the amniotic fluid. The mare will generally get up and down during this phase, trying to make herself more comfortable. Very soon afterward, you should see the foal's front legs appear, one slightly ahead of the other. This allows the shoulders to be narrower to pass through the pelvic arch. (This may be the reason why horses are either left or right handed.) Contrary to a longstanding myth, the foal does not have its legs joined together. Within five minutes, you should see the nose and then the head of the foal. At this point, you need to observe if the presentation is correct. There is no need to interfere with a correct presentation.

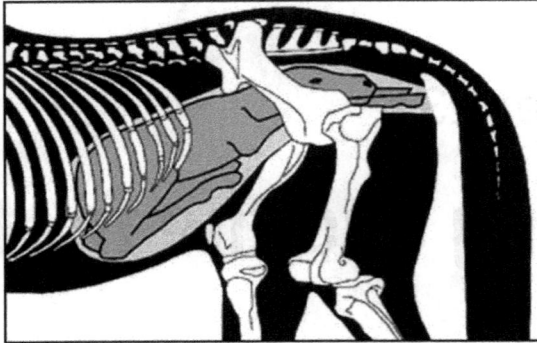

Normal position of the foal during foaling.
(From Evans, *The Horse*, 1990)

If the hooves are even, draw one more forward. If only one hoof is visible, it might have a leg folded back. This requires the attendant to glove up, push the foal back and insert an arm to unfold the leg and bring it back into the correct position. Do the same if the nose is not appearing.

Anything else you are unlikely to be able to solve on your own. While waiting for the vet to arrive a general rule of thumb for "dystocia", or foaling problems, is to resist the urge to try and correct it yourself if you are not sure what you should be doing, or if you've been unsuccessful so far. Ideally keep the mare walking until the vet arrives to help halt progression. Sometimes this can even send the foal back down the birth canal enough to correct itself. Difficult presentations require immediate assistance from a vet, with some requiring anaesthesia, a caesarean, or dissection of the foetus, called "fetotomy." Foals that have died in utero are often not presented correctly.

Another condition is the "red bag" delivery. The placenta has partially or completely separated from the lining of the uterus, prior to the foal being delivered. The reddish placenta, rather than the foal, is seen first. The foal will likely suffocate as it is no longer receiving oxygen. URGENCY is required. Rapidly cut the thick red bag with scissors or knife, tear away the placenta with your hands and give assistance with the delivery. This is the primary cause of a stillborn foal. (The other is infections).

The most difficult part for the mare is pushing the shoulders through the birth canal. She is usually lying down, giving very forceful pushes. Some assistance may be required if this seems to be taking longer or if the foal appears large. Clasp the foal by the forelegs, maintaining the uneven length, and pull down firmly in the direction of the foaling in time with the labour. Once the shoulders are through, the foal usually slides out quickly to the length of the pelvis while mum and foal take a break.

At this point, remove the sack from the nose of the foal if it has not broken.

If you are unfortunate to get a stillborn foal, the best thing you can do for the mare is remove it before she gets up. (Use a wheelbarrow). If not, you may have to leave it with her for some days till she walks away or provide her with an orphan to raise.

The mare will commonly nicker to the foal before she gets up and will stay lying down to regain her strength. Allow her to do this; otherwise, the umbilical cord may break prematurely. The umbilical cord transfers the final nourishment via blood flow to the foal while it is lying there, and it will break naturally when the foal gets up. Then spray or use a cotton ball with an iodine solution to swab the foal's umbilical stump. This is a recommended precaution against the nasty infection called Joint Ill. It is a very serious illness, tough to treat, and often results in the foal's death. Any foal born where others are regularly foaled may pick up something, even in a clean environment.

People who are foaling mares down regularly are advised to wear protective clothing. This is because quite a few infections can transmit to humans.

When the mare is standing, she will most likely lick the foal. If not, it is best to rub the foal with a towel to stimulate circulation. Allow her to do this, as it is her bonding time. It stimulates the production of the hormone oxytocin as she does so, which in turn stimulates milk let-down, contraction of the uterus, and expelling of the afterbirth. She may even chew on it. This is normal.

Allow the mare to lick the foal - Mare and foal of Brooke Allen. Photo by Emma English.

The afterbirth will normally slip out on its own when the mare gets up or within half an hour of foaling. I don't like waiting more than three to five hours because an infection can develop rapidly. Retained placentas are an emergency in horses. It is not advised to remove it yourself because you may unwittingly leave a torn piece inside. This removal procedure is best done manually by the vet combined with an oxytocin injection. In the meantime, if the membranes are down past the hocks, tie them in a knot to stop them from dragging or tearing, should the mare step on them.

Once you have the afterbirth, spread it out to ascertain if it looks normal in size and colour. You may notice a peculiar round, rubbery object, called the "hippomane". This is composed of fetal waste that accumulated during pregnancy. It is also possible to find a calcified ball, which is now recognised as being the calcified remnant of the egg sac, (rather than a mummified twin as was once believed).

A healthy placenta, hippomane and large yolk sac.

Next, check the mare for any abnormal tears or rupturing. She may have a haematoma or, worse, a prolapsed uterus. Both these occurrences are exceedingly rare in horses but need immediate attention. Another infrequent complication is the rupture of the bowel or anus, caused by the foal's legs penetrating the bowel during birth. Usually, it can be repaired with surgery, but it is best to retire the mare from breeding, as it can likely occur again due to the weakness of the bowel.

Over the next few days, watch out for pain, colic, bleeding, high temperature, fever, or other abnormal signs. Internal bleeding or rupture occurs more frequently in aged mares and usually results in the death of the mare and an orphaned foal.

A few mares, nearly always maidens, will make it difficult for the foal to suckle, swinging around or being restive. This is nearly always due to abdominal pain or super sensitivity of the milk bar. An occasional one might reject the foal. Very few won't respond to human intervention and patience. Some may be distracted by food. Others may require an oxytocin injection until a routine is established.

Watch out for the rare "foal proud" mare. If you have never had one, you won't realise just how dangerous they can be, but they can savage you, or pin you against a wall in an instant. Best to stay well clear of her for the first days or weeks until you can be sure she has settled. If you get such a mare, I wouldn't breed from her again.

The newborn foal

The foal is often up and standing before the mare is up. I like to see the foal drinking within two hours. Some are as quick as a few minutes. It may look skinny or "scrunched" up, but it will "un-scrunch" after a few days. Wobbly and bent legs are common and normally will straighten out after a few weeks.

A foal will usually drink between 250-500 ml. They drink little and often, about every 30 minutes. Failure to suckle is the first sign of something being wrong.

Madigan Squeeze on a newborn foal

If the foal makes no serious attempts to drink or appears to have no "suck reflex," it may be a "dummy foal." This is thought to result from a lack of oxygen during birth, a super quick or difficult birth, or placentitis in the mare. In this syndrome, foals appear detached, confused, unresponsive, or maladapted. There has been great success in treating them with the Madigan Squeeze, a technique that is easy to apply and replicates compression in the birth canal. It involves wrapping a foal's upper torso with rope loops and applying pressure for 20 minutes. The foal usually only requires one treatment. Learn about this technique used on newborns with dummy foal syndrome here: Video with Dr Madigan: https://youtube/IlVwUhwuHXk

The foal must receive the colostrum or first milk, which is slightly yellowish. Foals are born with zero antibodies, and only receive it from the mare from her colostrum. If it is having difficulty suckling, it may be necessary to milk the mare and manually feed it with a clean syringe. You can hand milk or use an "udder pump," which you can buy from most livestock suppliers. You can make one yourself by cutting off the end of a 60 cc plastic syringe with a sharp knife. Put the plunger through the cut end. Place the other end over the teat and gently pull down on the plunger.

It pays to have a supply of colostrum in your freezer, collected from several mares over the season. Place this in dated, separate small containers, (up to 300mls) as you don't want to thaw out the lot at once. You must slowly thaw it to room temperature, otherwise you will destroy the valuable antibodies it contains. You may need to feed several times over 24 hours before the foal is drinking on its own as after that there is full gut closure, and the foal can no longer absorb the antibodies. Very difficult foals will need to be stomach-tubed with a colostrum substitute by your vet or even be given intravenous plasma therapy.

Vital signs of the newborn:

- Temperature: 37.7-38.8 degrees C (99.5-102 degrees F).

- Pulse: 80-100 beats per minute.

- Respiration: 20-40 breaths per minute.

If the mare has been running milk before foaling or the foal looks weak, request an IgG test from your vet to see if it has absorbed enough antibodies. A normal level is greater than 800 mg/dl. Some studs test every foal routinely in the first 12 to 16 hours. I would suggest that new breeders have a post foaling mare and foal vet check with IgG test at 12-16 hours regardless. A foal is prone to hypothermia, as it cannot regulate its body temperature very well for the first few days of life.

Within a few hours after the foal has begun suckling, he should pass his first bowel movement, known as the meconium. This first movement is sticky orange-brown. It will progressively darken and become formed after a few days. If he has trouble passing it, you will notice him straining, with his tail up, and not being successful. He may experience considerable discomfort. A solution is to use an enema available from your pharmacy. The longer you leave it, the more likely he will not pass this on his own.

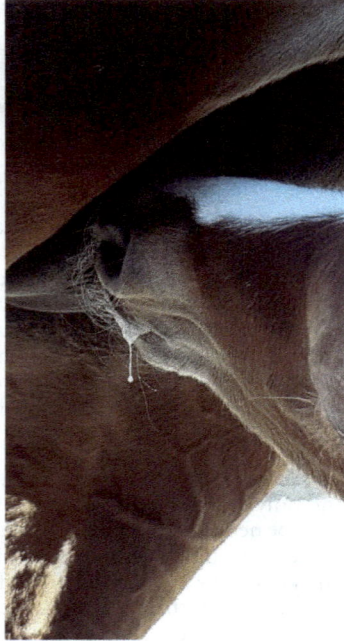

Photo courtesy Sally Gould-Hurst

On the other hand, you may notice he is scouring. Diarrhea in foals can be very serious. Due to water loss, they can quickly suffer from life-threatening dehydration. This may stem from parasites, or if from an infection it will present with an above-normal heart rate or temperature and requires veterinary intervention.

Scours often occur five to seven days after foaling. This used to be thought to be a response to when the mare comes into foal heat, but it is now linked to the foal exploring the tastes and foods of its environment at a few days old. The gut becomes swamped with its new microbes and can respond by scouring. The foal eating its mum's manure is vital because it is gaining the probiotics and gut enzymes that it needs to effectively digest these new foods. (Much like sick horses were often given a manure drench). Orphan foals still get "foal heat" scours. This is rarely life-threatening. Again it all comes down to a clean foaling environment as these scours are often not seen in immaculate grassy paddocks.

Premature foals require extra care. They are often too weak to stand up and suckle. Their immune systems are likely compromised, and they may need intravenous fluids. They may need a heat lamp in one corner of the stable. Rugging a foal is appropriate if the mare accepts this, but surprisingly, not all do. I have known a mare to attack her foal because it was rugged. It may be best to leave the mare and foal at a clinic with specialized expertise.

Foals born after a prolonged gestation are known as "dysmature". This condition is due to a poor intra-uterine growth environment which retards development. Therefore, the foal requires careful observation, similar to the premature foal.

Neonatal Isoerythrolysis (NI) foals, or previously known as jaundice foal syndrome (JFS,) is a condition that occurs when the foal and mare's blood types are incompatible. The mare develops antibodies toward her foal's blood type. It is most common in Thoroughbreds. After birth, the foal takes in these antibodies through the colostrum but quickly develops symptoms of lethargy, weakness, and anemia within 72 hours. If NI is suspected, muzzle or separate the foal immediately so it cannot drink from the mare and seek treatment. If caught early enough, the foal has a good chance of survival and can be returned to the mare after approximately 48 hours, when the foal's gut can no longer absorb antibodies. Otherwise, severely affected foals may need to be hospitalized for several days, especially if oxygen and blood transfusions are required. Mares can be screened to see if they will likely have future NI foals.

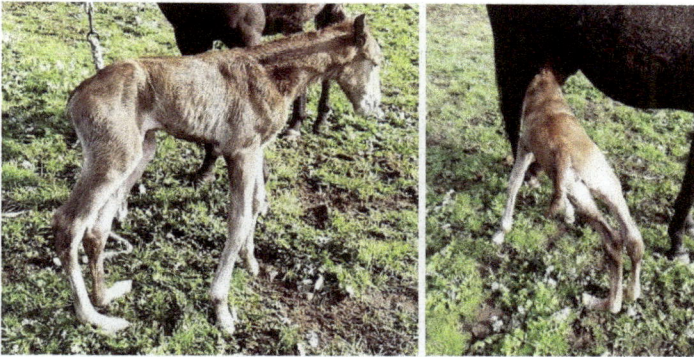

Windswept foal at 1 hour, below, at 4yo, Chalani Skywalk HSH.

The most common problem of the foal is bent limbs. Many will straighten on their own or by simply stabling/enclosing the mare and foal so that the foal cannot run around putting pressure on its joints, for up to ten weeks. If its tendons are contracted, such that it is unable to stand correctly, it will require splinting and bandaging. Splints can be painful. The foal may spend most of its time lying down, so provide soft bedding and painkillers. Be vigilant in avoiding pressure sores. Therefore, the bandages are usually changed every one to two days. Applying gentle massaging pressure three to five times daily on the joint in the direction you want it to straighten is also helpful.

Turned-out and turned-in limbs respond well to a simple rasp of the hoof every week or fortnight. You want to take off the outside of a turned-out foot or the inside of a turned-in foot. It is possible to fit caps with artificial glues to enable the corrective process in difficult cases, though this can be expensive. More difficult cases, such as the windswept foal or dropped pasterns (tendon laxity), are best followed up with a vet and farrier. This tends to occur in foals larger than expected for the size of the mare.

It is not unusual for a foal to have a hernia. Usually, this won't be evident for a few weeks as it takes this long for the navel to close naturally. Feel the hole through the skin. If it is more than two fingers' width, it most likely will not close on its own and should be monitored carefully. If any food becomes impacted in the section, it will be a hard lump, and the danger is strangulation of the bowel, which is likely fatal. In this event, seek urgent help. Hernias can be easily corrected with surgery, so it's best to get it done at three to four months if you haven't noticed it receding.

Worming of foals is not done before six weeks of age but check on the label as some medications should not be given to foals before six months. However, the usual vaccinations of tetanus, strangles, etc., can be given from six weeks. In Australia, young foals can be subject to heat stress. They go downhill quickly. Make sure there is adequate shade and water. Signs may include rapid shallow breathing, flared nostrils, staggered gait, and high body temperature. Never worm foals in hot weather, which can add to the stress.

The young foal will drink immediately upon awakening. Failure to do so is a red flag. Check to see if the mare's udder is full, a sure sign that it hasn't drunk for a while; if so take its temperature and contact your vet. Early action is best!

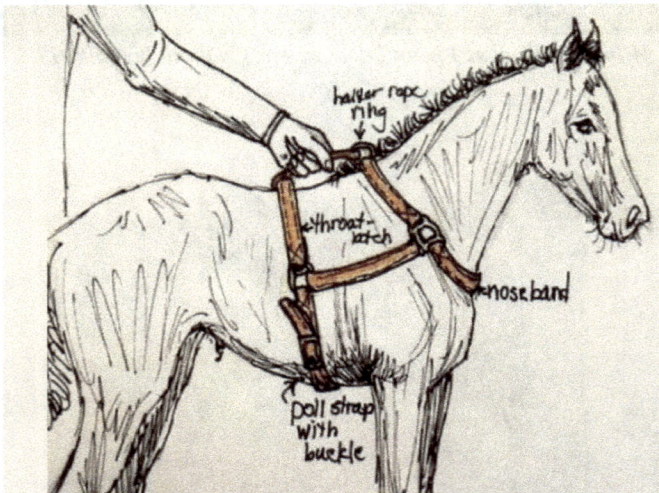

Suitcasing - for restraint or leading a young foal. It is easiest to use a halter with a buckle on the noseband so you have no need to slip it over the foal's head. Courtesy Red River Reproduction Clinic. Like their FB site for excellent information.

Arabian Riding Pony mare Glenvale Grace. Owned/photo by Maree Garratt. Below: Kyabra Park American Graffiti, Australian Pony. Owned / photo Tegan McKenzie.

Skyview Stud's Skyview Illusion HSH and foal, Skyview Infinity HSH. Photo by Jenni Phillips.

Chapter 24

Care of the orphan foal

Inevitably, if you have been breeding many foals, you will have an orphan. We've had two, one whose dam died of kidney failure at 10 days old. We immediately transferred it to drinking milk from a hand-held bowl and later a bucket. We then put the foal in with a quiet yearling. We soon discovered the yearling was also drinking out of the bucket!

The second orphan refused a bowl, and establishing any feeding routine was difficult. Another mare foaled a few days later, and as my daughter watched the birth, she made the snap decision to take the orphan out to her. She rubbed the placental fluids all over the foal. When the mare got up, she thought she had twins and bonded with both!

If a newborn, the orphan will need to receive a replacement colostrum and/or plasma infusion. This is important for it to thrive and receive protective antibodies. The foster mare may provide this if she has lost her foal in the past 24 hours, otherwise, draw from your frozen storage, (slowly bringing it to room temperature). A vet can give this by stomach tube if necessary. An IgG test must be given under these circumstances.

Horses need *horse* antibodies, so bovine colostrum powder doesn't help them achieve their IgG levels. That's why it's important to have stored colostrum or go the plasma route (which can also be given orally in the first 24 hours but expensive.) USA readers can use Seramune which is an injection or oral paste for newborns that replaces colostrum, but not available in Australia. Give plasma *before* feeding milk replacer or they can have a bad reaction to the bovine protein they absorb in the milk replacer in the first 24 hours.

"I share this story to give others the courage to try this if it becomes necessary. My favourite mare had cancer. The vet told me I had limited time. So I thought I would try for one more foal as her last one died, and I never bred her much because I always wanted to ride her. Angel went in foal like the good girl she had always been, but the cancer progressed quicker than we thought. By the end of her pregnancy, it was obvious she would not make another six months.

"I began planning the switchover from Angel to her best friend Janara Glamour. Glamour had a foal the previous year, but her foal had been weaned. A 10-day course of hormone treatment was given to Glamour to induce milk production. I started milking her three times daily from day four. Over the next six days, I worried that Angel would be distressed, Glamour would refuse the foal, the foal would get hurt, etc. However, none of those things happened.

"The vet, my daughter, and I were amazed at how calmly Angel accepted everything and how smoothly everything went. Angel has left Glamour and me to raise her son. And, if he is anything like his mother, he is going to be amazing". – Janelle Groeneveld

Most studs do a "call-out" for a foster mare which will have lost her foal or been brought into lactation via veterinary intervention. The introduction needs to be done carefully. It is best to match the mare's size and stage of lactation to that of the dam of the orphan. The foster mare may be "tricked" by dousing her own urine and manure copiously over the foal, even blindfolding her. She may require sedation. If no mare is available, a milking goat standing on a platform has been successful.

If you rear it yourself, feed every three to four hours with a milk replacer formula in a warmed bottle with lamb teat, or a bowl. The digestive tract of a very young foal is undeveloped and requires small feeds often. Otherwise, it will likely develop scours. Its digestive tract matures as its dietary requirements change. Initially, it must be fed at least six to eight times daily. If it develops a "pot belly," you are probably not feeding little and often enough. When eating well, you can substitute powder with milk replacer pellets. (Not available in Aust).

The orphan will need company, even if it can see another horse next door in a stable, otherwise, you are setting yourself up for handling complications and "brat" behaviour. From two weeks onwards, as they start to get teeth, you can offer very small quantities of free choice feed, such as good quality grass hay, leafy lucerne (they will pick at the leaves) and pellets. Feed so there is always a little bit over when you check the next time. That way, you know you are keeping up with its appetite. You can wean off milk at the start of the third month, though longer is better.

Chalani Laura with her natural (brown) and adopted foal (bay). Photo by Kim Ide.

Chapter 25

Raising foals

Training and weaning

It is important to have a toolbox of different techniques depending on the age and amount of handling the youngster has already received. This will also depend on your facilities and your staffing. But why should any horse be happy to be caught? It often results in lessons, being ridden (hot, sticky, worn out, sore mouth, etc.) standing tied up for long periods, or a float ride, or the vet or farrier.

The obvious solution is to catch the horse multiple times where none of these events will happen. This should occur on your daily health and safety checks. Take carrots or a bucket of feed. It starts with being around your horses as foals. Just walk in with your horses and establish a friendly atmosphere regularly. Every time they come in, particularly with sensitive and hard-to-catch horses, let them stand with feed in front of them. Feed can be a distraction, attraction, or relaxation time. Any horse that refuses to eat when tied up is stressed and needs frequent opportunities to be brought in to learn to relax. Better still, have a mate eating alongside.

Never leave a halter on a horse because it can't be caught, especially if you are not checking on it daily. Halters unsupervised are one of the most dangerous items for a horse. Just ask vets what injuries they see because a halter has been left on. I have seen youngsters with heads growing into halters and being unable to chew properly because of bad practices. If you must leave something on a horse that is hard to catch, use a neck strap up near the jaw or a special breakaway head collar. You can even dangle a length of plaited twine from it.

How to put a halter on correctly

Easy right? Many do it incorrectly, even people who have been in horses for years. It is a practical test I would give if I considered employing someone. The way they perform this simple test tells me so much. The correct technique must be a habit to avoid injuries and do the job efficiently and without drama. It should be taught at Pony Clubs and horse-riding establishments. So here it is. Approach the shoulder, not the head. Note: the nearest hand (left) goes *under* the neck, transferring the end of the halter to the right hand over the neck on the opposite side, while holding the nosepiece open in the left hand, to allow the horse to put its nose in by itself. Ensuring you know how to put a halter on correctly, is quickest, safest, and minimises the chance of the horse breaking away from you when you catch it.

Another trick is to feed out of a nosebag. The youngster gets used to having something placed over its ears and head. You can soon add a halter. If this or the bucket approach doesn't work, try bringing in another horse and have it follow that horse. It won't take any longer.

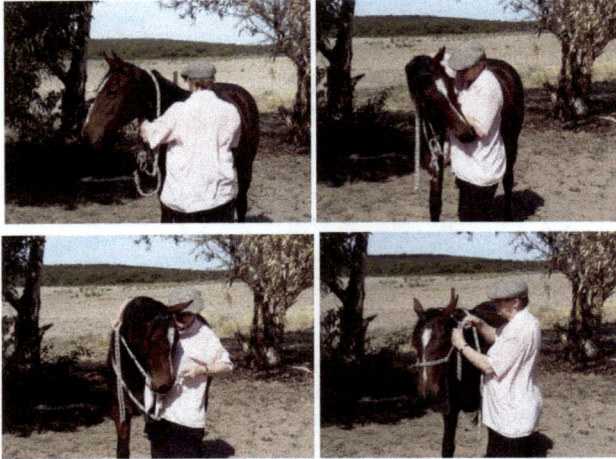

How to put a halter on a horse

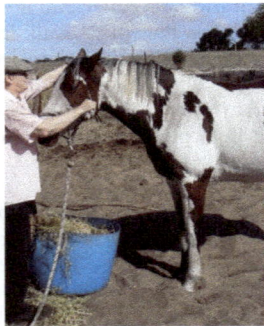

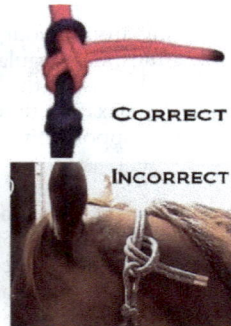

CORRECT

INCORRECT

How not to do it.

Wear clothes with a belt to lead something with it in an emergency. Never chase a horse that won't be caught, flap your arms, or create extra chaos. And don't click to them! These just reinforce problems and increase your frustration level. Be patient and allow for plenty of time. Plan ahead.

A set of yards is the only option for the serial hard-to-catch horse. Usually, a chute, or gradually narrowing area to the yard, is needed. It will be more enticing if horses are regularly fed in this yard. Leave the horse there to eat until it is more relaxed. Don't just pounce, catch, and walk it away without reward.

I have known older horses to run around in the yard as if they are free lunging, then all you do is say "whoa," and they stop. Teach it to be stroked with a lunge whip, (up and down like a violin bow), then bring the whip higher until it is on the horse's neck, still rubbing up and down. Quietly drop the lash to the other side of the horse. Close in a little until you can clasp the end of the lash, from under the neck. If needed, you can tie the lash to the ring of a rope, to draw back over the neck.

You may make a squeeze between the difficult horse and another. You may ride in on a horse, squeeze it up to the fence, and put a rope over its neck. Kel Jeffery used a pole with a rope noose at the end. You may have other tricks up your sleeve for a particular horse.

Catching the foal

When you want to catch a timid foal for the first few times, it pays to squeeze it between the mare and the wall of a stable or yard. Tie the mare up. Stand someone at the mare's head so it doesn't duck under. Then, walk up to the back of the foal and slowly slide toward its shoulder. At this age, you should be able to clasp the foal around its shoulders and hindquarters like a little cage without touching it. Next, bob down and make yourself smaller, encouraging it to sniff you. If it tries to run away, hold it firmly until it ceases to struggle, and give it plenty of time to relax as you quietly stroke it. Most foals accept touch if you can find an itchy spot and it may even attempt to scratch you back.

Turn the foal around, so it faces you and allows you to stroke it from the front on its neck. If it resists this, place its hind end in the corner. Most people make the mistake of trying to stroke the foal's face first, and it jumps away. Allow it to sniff you during breaks.

I like to have a little rope around the foal's neck and a rump rope to start leading off in a circle around me. This will encourage forward movement before I put on the halter. It makes haltering so much easier if it responds to the neck rope before a halter. It means you can stop and turn without pulling on the head. Some foals will react to pressure on their heads by running backward or even flipping over, so it is best to take this step first.

Forward foals and those with some early handling can have the halter on immediately. Many foals accept the lead up but react as you put the halter on. Just put it on anyway and walk away. It will soon get used to it. The final step is to attach the lead rope and teach the foal, while standing still, to bend its head towards you both ways. Additionally, I like to pass the rope behind the foal's rump on the opposite side, bending the head around until it has turned full circle to face you. Until it will do this softly, it is best not to attempt to teach it to lead from the halter. Continue to use the rump rope for "forward" right up until weaning if necessary.

Each session should be short, with you stepping back from the foal after a few minutes, then repeat after it has relaxed (chewed) on the other side. Using advance and retreat, begin rubbing the legs and picking up the feet. If the foal kneels when you touch its legs or shows ticklishness, pick the leg up from the fetlock joint instead, then when the leg is held up, stroke up and down the leg.

The feet are held up for short times until there is some relaxation and promptly placed on the ground again. Ask it to wait for the leg to be put down rather than pulling away from you. Learning happens on the release of pressure. That means if you don't release the pressure, the horse will not learn. You can encourage relaxation by circling the foot while it is in the air, holding the leg higher, and supporting the foal under the neck with your free hand, so it doesn't overbalance. Do both front and back legs. If it struggles while you are doing any leg, go with it and hold the leg higher by flexing the hoof more until it balances itself, using the wall on one side to limit its movement, then release. Put the foot down, rather than the foal taking it from you.

The final step is to walk the foal, with the foal walking beside you. You can start by leading the mare in front of the foal, or signalling on the rump rope before asking with the halter. Your left hand is on the lead, while your right hand is holding the rump rope over its back, so that you can give it slight tugs to "come-along." Others use the method of bending the head around slightly so that it steps across several times. After that, it will start to follow the head and take bigger steps. Either method works quite well. Find out what works best for that particular foal. Always go back over previous steps before progressing to a new step.

Chalani Ripple ASH at 3 days. The rope goes through the noseband, not the ring.

Teaching to lead is not that difficult if done consistently. You don't want to get into a dragging match, which causes damage to the neck and teaches nothing. Go back to bending the head around if the foal starts to drag and give it a few tugs with the rump rope to sharpen it up if necessary. Take the foal away from mum, by degrees, to a distance both are comfortable with. Gradually increase the distance over time. Give the occasional "back up" with pressure on the chest, just at the groove of the neck, followed by slight pressure on the nose. This is the first step in teaching the horse to stop from slight nose pressure. However, the aim it to have it stop with your body, before it needs nose pressure, achievable if you keep a slack hold on the lead. The biggest mistake novices make is to have a constant hold on the lead rope, which deadens the foal, rather than keeping him soft.

For most foals, this can all be achieved within a week of short daily sessions, but they tend to regress again if not repeated a few times a month (which we do when feeding out). We never owned a stable for years and did all this in the corner of a paddock.

For the hard-to-catch foal that has already had a few lessons, squeeze it between the mare and a wall. Hang a long rope with a ring on the end over the foal, from across the back of the mare. It will either shoot forward or back. Either way, it should carry the rope with it. You can then pick up the rope from a distance, slide it through the ring, and position it up the neck. A pull and release at *only* 90 degrees to the neck will change its mind, so you can walk up slowly and put the halter on.

Youngsters in safe yards such as a lunging ring, may have a halter and a long thick drag rope during the day so long as they are under observation. Do this for a few days in succession. It encourages them to stop when they step on it and ignore it when it drags along. This is a really effective pressure and release technique that directly connects to the horse's mind. The youngster learns to think before it reacts to undo the pressure it has created by stepping on the rope.

If you use either method, make sure you put a halter on as you catch the foal. If not, as soon as you take the catching rope off, it still hasn't learned to be haltered, and you are back to square one.

Larger operations have methods whereby an *older* foal is tied to the mare, and the mare is led around. This is after it has learned to give to the head.

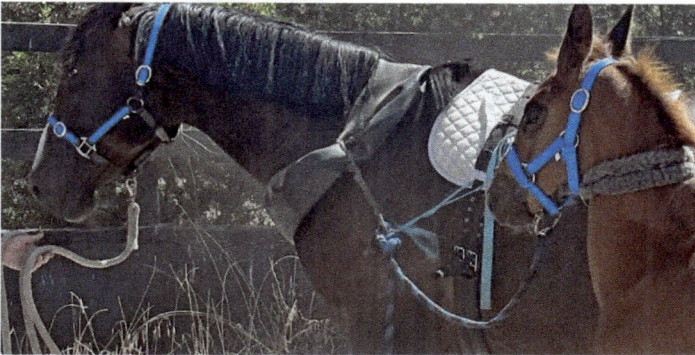

The mare has a soft rubber inner tube around its neck, and led so it can be stopped at any time. Photo from Dr Kerry Mack.

You may give a first float ride before weaning simply by leading it alongside its dam into the float. Or you may need to "wheelbarrow" it into the float by one person at the head and two people locking hands behind the foal, lifting its back feet slightly off the ground, and pushing it in by the front legs. If you need to load a larger foal, say to head off to the vet, it is a very useful technique.

Some studs prefer to handle their babies only *after* weaning. For routine procedures like worming before weaning, you need better facilities to do it this way and usually more expertise handling unhandled horses. I feel that the better approach is to have them handled early, and I know it sells horses if they are.

"For many years we have done this with weaners who have not been handled while on their mothers. First catch your weaner, by getting him to stand still and be approached, then touched with a long thin poly and then remain still while a halter is fitted. Surprising how simple this can be if you read your weaner and at the first sign of worry, just back off and come in again. On occasion we will rope one who is really "thingy", let him scoot around the yard with no pressure until he stops. Let him think about it a bit then start the approach. Same thing to back off when he gets worried but put no pressure on the rope. Just show him to stand of his own volition until one can go in, touch him briefly on the neck a time or two and fit a halter.

"If he takes off, just wait for him to stop then come in again. Never get concerned if he gets tangled up in rope. It is a very detoxing process once he knows he can get out of it himself. You might need to let the rope go and then pick it up again if he gets in a real tangle, but no

pressure on the neck. We always use an ordinary rope halter with a bit of feel to it (not a floppy, soft thing), and a very long lead rope tied to the halter, with none of those swinging, distracting snap hooks." - Merrie Elliott.

A good broodmare will take no nonsense from her own foal and if raised with others, it will have learned socialisation skills, language and boundaries. But you are the one who must follow on, by being fair but firm. There is nothing better than a well-mannered horse. It saves a lot of costs if you can do it yourself. Manners ensure the reputation of a stud, and it starts with the foal.

Weaning

Weaning is considered the most dangerous time of a horse's life due to stress, chances of injury, and illness. If you have given your foals some access to hard feed and handling, the weaning process will cause little fuss. Indeed, you will be surprised at how simple it is. The industry standard is six months of age, but you can do it early or later, depending on preference. We like all our foals to be born at the beginning of the season so they enjoy spring grass; they are all weaned around the same time, and we can put more feed into them if required.

There are several effective methods. One is to have all your mares and foals in a close paddock, then gradually take each mare away, one at a time to the back paddock, preferably out of earshot. There may be some calling out, but it is usually finished in a few days, and the foal remains comfortable with his mates.

The second method is to have yards inside the weaning paddock. Separate mare and foal for feeding to get it used to the yard. Next, leave the foal in the yard (or you can do it the other way around if you prefer.) The mare is free to roam as she pleases and is only brought into the yard for the foal to have time to drink. After a few days, the foal remains in the yard, and the mare can go back to her original paddock, while the foal remains with its other weaner friends. Before removing the mare, take a little time to handle the foal, especially ensuring you can rub it while it eats its hard feed.

Double-check your mare's udder for a few days, and milk her a little if she appears uncomfortable. If it feels extra hot, she may be coming down with mastitis, which, although uncommon in mares, can happen, and without treatment, she can become very sick.

If you only have one foal, try weaning him with a gentle pony or older retired horse. Weaners need to be separated from their dams for six weeks, (some need longer), or they will induce the mare to lactate again.

Training the weanling

It is time to lead with a longer lead rope and introduce the dressage whip (poly pipe / flag / stick) for the "go-forward" lesson. We carry a whip and gradually dispense with the rump rope. If you don't carry one, you may need to flick with the end of the lead rope. An extendable "teacher's cane" can be bought online for about $10. The horse needs to learn the difference between "go forward" and "come forward," depending on where you stand.

A youngster may lead well but not stop. Take a rope training halter and give a stronger tug/release on his nose. You do not want him to circle you if that is the only way you can stop him. Wave a whip or poly pipe in front of his face like a windscreen wiper and let him run into it across his nose if necessary. Then ask him to back up to the point where you *first asked him to stop*, using the end of the poly pipe to push into his chest. Repeat this each time so he listens to you. This also has the benefit of teaching him to stand still.

A weanling that tries to run through you can have a neck hold around the gullet or a karate "chop" under the neck. There is no excuse for a youngster to push into you with its head or by bulging its shoulder. Teaching the horse to *turn away* from you from the slightest touch, using a stick tapping on his neck, will generally solve this problem. Or, carry the end of a brush near your hip so that it hits the brush any time it runs into you. It needs to be something noticeable. The youngster knows the fence doesn't hurt it unless he runs into it. Teach him that you are the same as a fence.

A minimum routine is to teach the weaner to lead out at walk alongside your shoulder, tie up, pick up its feet, load into a float, and be taken for a float ride.

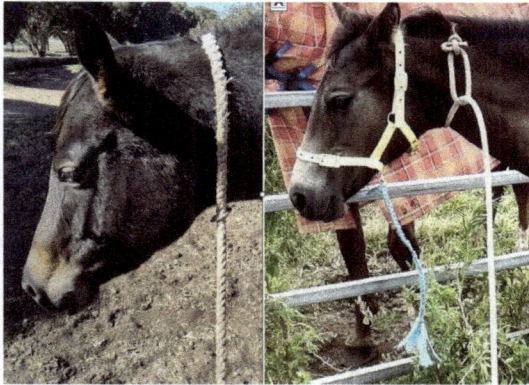

Rope with a ring. LHS: correct position at the base of the throat so it will loosen when the rope is dropped. RHS: incorrect. Then put the halter on OR thread the rope across the nose into the loop around the neck for a makeshift halter. (That is known as a "war-bridle" or a "come-along.")

Teaching to tie up

Tying up is just a natural extension of everything which has come before. Every time you hold the foal while he is standing is, in essence, a tie-up session. Sit on a tyre or log and let him stand there for a while. Then, each time he moves, bring him back to the same spot again with a turn of the head, the same as what would happen if he was tied up. We don't tie up solid until after a foal is weaned and only once he answers "come forward" 100% of the time. Tying up then, is a non-event.

First, practice "tying up" with a long rope wrapped around a pole, with you holding the other end a distance away. You can easily release it if you need to.

Next, put a long rump rope on because this encourages him to move forwards *if* he leans back. The rump rope is threaded through the halter. Take a neck collar that is weanling size and tie the lead through the nose of the halter as well, to something solid *at a height just above his withers*. The wide neck strap spreads any pressure as opposed to a halter with no spread. This ensures he can relax while tied, but should he pull back, the collar is at the right height not to damage his neck. The rump rope should be tied short enough that it is the first rope he feels, and he immediately comes forward again, and secondly the neck collar *if* he should pull back harder.

Always use a quick-release knot and be nearby to watch. Tie for short periods at first, gradually increasing the time.

You can tie to a tree, a pole with tyres around it, or a rubber tyre ring set up for the purpose, which allows a little stretch, or the top of a pole or barn roof. The footing should be soft because they can skid their back legs if they really go to town. Few will, though some will put on a "tantrum" of sorts such as pawing, before they realize they are there to stay. A particularly restive or nippy horse may be cross-tied if you have a suitable facility.

Yards, where they are likely to put their legs in the rails, are unsuitable. Tie-up rails are usually too low for training purposes. Never tie up with a thin training halter. They are for leading only. Horse chiropractors treat many neck injuries caused by horses pulling back when tied up solidly. If you doubt the safety of tying up solid in a particular location, tie to a loop of string (baling twine). And never deliberately frighten them in order to "teach them never to pull back!"

We like giving them a hay bag once they are a little more settled, so it becomes part of their routine. The final step is to tie them up both outside and inside a float with a feeder, with a friend eating nearby. Never tie up with the rope long enough to put their leg over or head under it.

Another method is to tie up with the front leg to a hobble strap. Advocates of this system say it is a much easier, gentler method that puts minimal pressure on the neck. It is taught once they lead and know how to come forward. The beauty of it is the seesaw action, so when they go back, the front foot automatically comes forward, which means the horse then needs to come forward to put its foot back on the ground. The near front is done the first day, followed by the off front the next day, to keep them even, then alternating until they have learned and are okay to tie up normally.

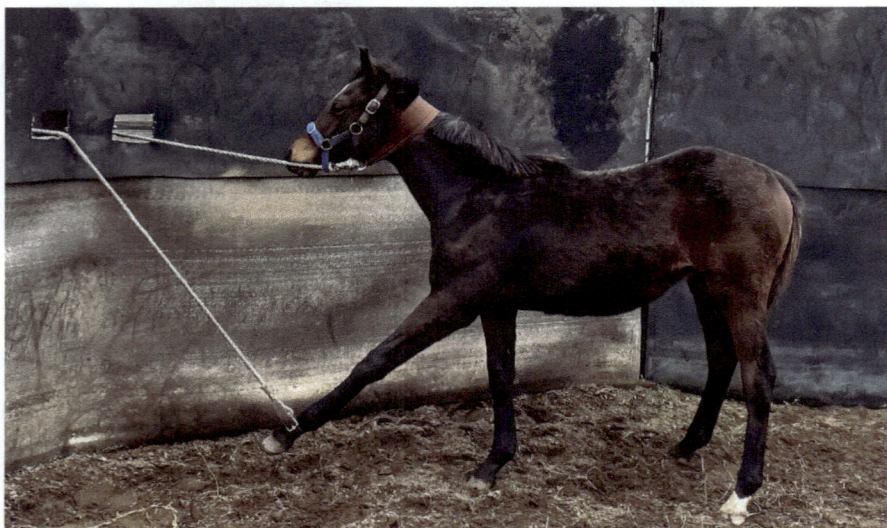

Note the full rubber wall and soft sand surface. Photo Peter Haydon

Chapter 26

Further training

What should the young horse know?

Following weaning, the very least a young horse should know are: lead, tie up, eat from a hay bag, pick up feet, load onto a float, and have had a float ride, before being sent off to a new owner. Always use a nice long lead rope, with a spliced end so that it has "feel" to it. Without the nub at the end, it is easy for the lead to pull through your hand. Without length, you are likely to find yourself in a bad position because you can't let the rope out.

Hosing and washing

This is one of the best educations you can give a youngster. He learns acceptance; even if he absolutely hates it at first, you can just follow him around until he gives up. Start quietly on his legs before moving up to his body. Don't hose on a cold day, unless you have warm water facilities and he has already been taught to be rugged. Scrape him off. Be considerate, but don't put up with nonsense either. Continue by ignoring it and just keep going until he tolerates the water. Don't do his neck and head until he masters the first few times, then only quietly with a gentle flow.

Rugging

The question of rugging or over-rugging is mostly an Australian issue because the rest of the world uses stables and barns more often for shelter. Stabling horses is labour intensive, and adds to costs, besides being unnatural for horses. We have land availability and a culture of keeping horses outside with New Zealand rugs. Make sure you choose a well-fitting brand, with sturdy snaps and straps. Rugs will assist an older horse to maintain weight. Even show horses may be maintained 24 hours a day with a rugging regime. We showed our horses for years without ever owning a stable.

This presents a problem for those who work off site, to rug/unrug in accordance with the weather. We like to use the shade-cloth rugs in warm weather, which allow airflow and protection from sun bleach and insects, but allow the horse's coat to breathe and stay clean. If the rug gets wet, it doesn't matter. Horses do well in these rugs, and due to its protection, will stay out feeding longer.

If teaching a youngster for the first time, be aware he may run from the leg straps, so put it on in an enclosed space, such as a lunging ring, in the first instance and let him wander around in it. Others may accept it quite happily until you go to take it off, so tie the horse up, the first few times.

Hobbling and stockwhip

Hobbling is excellent for training horses to stand still if they become entangled in wire when they might otherwise panic and injure themselves severely. It is useful on stallions that might run their fences and lose weight, or horses that paw when tied up. They must be taught in a soft area first. It is good to teach them early on as an extension of the tying-up process, or ask your breaker to do it. Some breakers will do it as a matter of course, so always check. You can hobble older horses which are hard to catch with either a grazing hobble or sideline. This is done when driving cattle, so horses can graze overnight and be found the next day. Watch an "old hand" do it before you try it yourself. You can also crack a stockwhip around him. This is an excellent method to teach a horse to stand still.

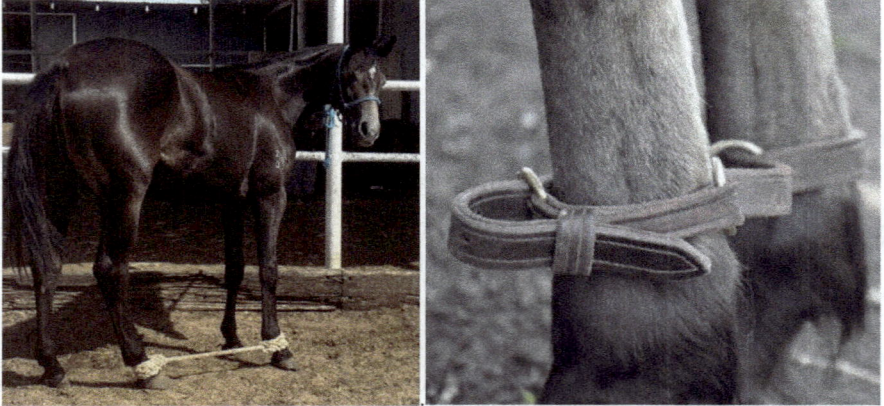

Two types of hobbles. Left: sideline. Right: breaker hobbles. (Keystone equine)

Float loading

If they respond to the head with the "come forward" and "go forward," they will lead almost anywhere. Lead them in and around obstacles, over tarps and concrete, into your shed, over rails, up steps, through water, over tyres and bridges, and anything else you can think of. The final obstacle is the float. Teaching a horse to go into a float is one of the simplest things if your horse will lead properly.

First, secure the float to the car! Remind him to go forward with the rump rope if he is reluctant after a few tries. Take him in and out a few times, always rewarding him with food when inside. If you only have one weaner, pop a quiet pony in with it for the journey. Floating him around the block is usually enough for the first time. Most take it very casually, but if you suspect he may jump over the front bar of the float, then take a second tie rope under the front chest rail and to the tie-up ring so the youngster cannot raise its head more than a comfortable height. We have a short tie-up rope with a quick-release clasp at the end of it. In an emergency, you can then release the horse without climbing inside. Our horses are booted up in the float and always given hay. Never ride in the float with a horse.

When he is comfortable doing this, goes in and out easily, and has had a short ride or two, start teaching him to self-load. You will need a whip or poly-pipe in most instances to teach him and a longer lead rope. We use the Tom Roberts method, which is excellent for its simplicity. The principle is not to allow the horse's head to move anywhere except pointing into the float and never take him away or circle him to reposition him. He can step into the float from any angle if you keep his head pointing to the float. In the beginning, this may require two people. The person at the front should be back far enough so that the horse is not confused by his presence. The active person is the one who has the whip. The idea is to use the whip only as an irritation by flicking it on his flanks or hind quarters. Horses feel flies, so your horse will feel this. It will irritate him eventually so that he will move - sideways, forward, or backward. Unless it is forward, keep tapping by going with him. But if he does move forward, even by only a half step, stop immediately. If he puts his head down to sniff the tailboard, that is forward. Stop immediately. Wait for him to chew. Ask again, repeat. As he has already been in the float a few times, he will likely have no problem with this.

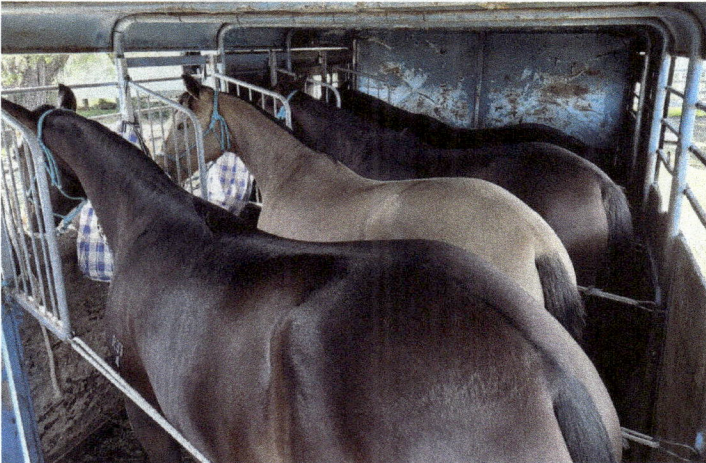

Braeview Stud youngsters having a practice float ride. Photo Jen Redgen.

The final step is to remove the second person, and you lead him straight up the tail board yourself, stepping aside to tap him with the stick to send him in the last little bit if necessary. I have retrained many a tough to load problem horse by this method. It works.

Never allow the horse to run out when the tailgate is dropped. *For safety, always undo the rope first.* He should wait until you ask him to back out with a tug on the tail and a tug on the lead rope if necessary. If you don't do this, he'll learn that the signal of the tail bar being undone is the signal to exit, setting yourself up for the horse which rushes out. Angle load floats don't have this difficulty; however, the horse may still turn and rush out before you are ready. Better to avoid this by backing him out.

Always allow yourself enough time to complete the training. If you are in a rush, it will backfire. The first time you head off to an event or show should not be the first time to float-train them.

As a yearling, a buyer can expect that a youngster will lead fluently, will trot out, and have been exposed to obstacles, a progression of his experience as a weaner. He should be good for the farrier and know to tie up, be brushed, hosed down, and possibly rugged. As a late yearling, he can be lunged and led off another horse (ponying). We like to take yearlings with our show horses for exposure and to stand at the float. Always use a long lead rope when training, and preferably gloves.

Tying the front leg up

Using a stirrup leather, or hobble strap, you can tie a front leg up, from a few weeks old. This is particularly useful if the vet comes, as a SAFETY measure, for the horse stands and cannot kick properly being on three legs. It is an old and proven method as shown in the photo attached of a mare being milked in Kyrgyzstan. It can be the first thing you do if you have to treat a horse or to do any of the multiple tasks required. Especially for inexperienced handlers, it eliminates so much risk. It is also helpful to retrain a horse for the farrier.

Breed shows are an excellent experience for the young horse. Showing in hand can promote your name and give your horses good training. Considerable training time is necessary for both standing still and standing up for the judge (posing) in formal position. You can accustom them to radio/music in the stable. If the work is done at home, showing is not difficult. Otherwise, a lot more exposure is required. We don't show without first teaching a horse to lunge and be bitted up (if required) for the class.

I do not like clipping manes off horses just for their "training." You can simply turn the clippers on and let them feel the vibrations. And certainly not one's broodmares, or horse out in paddocks. I feel the same about clipping foals' heads and necks for photos. These are welfare issues. Ugh!

Boundaries

From the time he is a foal, he is learning from you. Don't be afraid to set boundaries. After all, he learns them from his dam and while socializing with others. He watches body language and learns via cues that each horse gives, such as flattening of the ears or kicking or rearing, the horse's natural language and play. On the other hand, humans have flattened ears and are always rearing and reaching out with their front legs! Too often, people vocalise meaningless prattle or are inconsistently clicking and clucking, leading the foal to feel threatened or confused.

We don't feed foals from our hands, only from a bucket. If a foal starts nipping or pulling at clothing, it might be funny the first few times but can soon become dangerous. A gentle mouthing will receive a gentle poke; a sharp nip will have a sharp slap on the nose, or anywhere in reach, such as his ribs. Just enough for him to take notice. It should match the degree of wrongdoing. You can't hurt him more than the others in his paddock. Yes, he might pull away, so make friends with him again.

A foal that charges must be bopped by a piece of poly pipe, struck with a lunging whip or pulled up with a Jeffery rope. Some owners do not keep boundaries, especially concerning children or dogs. As a result, the foal becomes annoyed or sees the child as a plaything. Be extremely careful of any horse around children, and never allow them to be unsupervised. Another bad behaviour is allowing a foal to turn its hindquarters into you for a scratch. *Why allow something that you must later unteach?*

Never be angry with your horse – it is a sign you don't have the tools to go further. Stop. They unlearn as quickly as they learn. Many times you can restart the next day and not have a problem. Beware of behaviours such as fidgeting or pushing that you inadvertently allow at home, then expect them to get right elsewhere. Be consistent.

"Horses are always speaking to us. It's our job to be aware enough to listen and interpret without emotionally reacting. So much of working with horses calls on us to put our ego aside and instead build a relationship based on trust and respect (developed through consistent boundaries - which build consistent expectations) and a new-found understanding and connection.

"To create a respectful partnership, we bring boundary setting into it. Setting healthy boundaries leads to safety, certainty, trust, and respect within the relationship. Loving on a horse without teaching the horse to respect boundaries results in dangerous behaviours from a 600kg prey animal.

"We also need to be able to adapt our communication styles and be emotionally fit, to work with horses. Through our sound leadership, we help equip our horses mentally and emotionally to become the consistent leader they can trust. We don't help horses develop self-confidence or become sound decision-makers by trying to control every outcome. The only thing that is certain (after death) in this world is uncertainty. So, we lead them in a way that helps equip them to manage themselves in uncertainty. We can't rob them of their developmental experiences by trying to control all the outcomes and doing everything for them when it comes to training a horse. We have to train a prey animal's brain (which is completely hard-wired to flight or fight and self-preservation) to become self-sufficient and be able to stay calm in an unsafe or uncertain event.

"They need to have the space to make mistakes and be guided back to a place where they can try for a sound decision again. How we handle their mistakes contributes hugely to them building confidence in themselves or shutting down or bursting out in undesirable behaviours. Are we still talking about horses? Training horses is parallel to parenting children." – Jess Keenan, Leading Change Experiences.

A youngster won't learn boundaries if you do nothing. Take him places. Earn confidence and trust by remaining calm and consistent throughout his training. Correct him. Let him learn the rules. All of this is done to make your horses pleasurable and safe to be around. Don't expect the breaker to do all the work for you. It will take longer and cost more, and when you get the horse home, it will fall back into its old habits.

"Just as children aren't cognitively capable of seeing the end result of most decisions and behaviors, neither can our animals. It's incumbent on those around them who can see that path and those possibilities, to help them develop the tools and wisdom to deal with what happens when things get difficult. If we fail to do that, we're failing to shoulder the responsibility of stewardship, which should not be taken lightly.

So, the discomfort of having my horse or my child get upset with me because I'm not asking their permission, but rather insisting on obedience, is far less than the discomfort of watching them experience mental or physical anguish that proper preparation could have prevented. I can listen to and acknowledge their emotions and concerns while still honoring the responsibility of stewardship. Insisting on obedience and compliance doesn't have to be harsh or unfeeling, but it does have to *be.*"- Adam Till.

Have fun educating your horses - Braeview Sterling HSH yearling with Dan Redgen. Photo by Jen Redgen.

A horse is an accident waiting to happen. An emergency is the wrong time to train your horse.

Chapter 27

Starting under saddle

So far as the horse is concerned, starting is merely a progression of the groundwork given previously. Always go over the groundwork first. We will *never* back a horse before it is two years old, and sometimes not until it is three. This allows time for maturity, especially as the bones of the horse, even in breeds thought to be more mature at a younger age, like Thoroughbreds and Quarter Horses, do not finish maturing before they are five years old. Even our Futurity horses are not broken before they are two years old. I would never knowingly buy one that has been broken earlier, if only out of protest.

If you are sending the horse to a breaker, ask a lot of questions, and preferably obtain references from people who regularly send their horses to him. You may need to book well ahead. Do not decide on price alone. Visit to see facilities and the breaker working with his horses. Does he work empathically or grind down on the horse? What is included in the breaking process? What extras can be included? Are the facilities in good order, and do they appear safe? Does he use splint boots? What does he feed? If the horse will be put through a sale later as a ridden horse, will the trainer be the same person as the breaker?

Don't send a weak or immature youngster and do check if the young horse has wolf teeth or caps. These should be removed before proceeding.

You don't want the horse to come back underweight, scared, or "ruined". If the process is too fast, it may come back confused, fatigued, shut down, head shy or "overbent", a term used for a horse which ducks its nose low and behind the vertical, such that it is avoiding the bit. It may have experienced pain from girth galls, rubbed mouth, footsoreness or splints forming. If this is the first time the horse is off the place, it may take it a while to settle in to eating and drinking. This is a sure sign of stress, which can be alleviated by your preparation. You owe it to the horse. You don't want the horse to come back hating its experience or with problems he never had.

Most trainers appreciate the horse which has excellent ground manners. Don't be fooled into thinking because the horse is "good" at home, it has good ground manners. It may simply have become routined to your way of doing things. The breaker may need to start all over again to teach it some manners. In this case, the breaker may prefer less handling, than to have to undo your bad habits with the horse. Breakers need to consider the follow-up work the horse will get once it leaves his place, its future use, and the general ability of the rider. The ridden education of the horse merely starts at breaking. It should never stop there. Make sure you have a couple of trial rides on the horse at the trainer's place, before bringing it home. If this is not possible, ask for video.

I do warn against spelling the horse for too long after breaking if you're to ride it yourself. If you consistently find your horses are sitting in the paddock doing nothing for long periods, ask yourself, whether it is fear or lack of passion holding you back. Don't fall into the trap of breeding more, while you're doing nothing with your 3-year-olds, 4-year-olds, and the like. They don't sell themselves, especially if you are not educating them or sending them off to be educated further. Buyers want to see a broken horse ridden. If this is not you, re-evaluate if you should be breeding at all. It doesn't help you or your horses to let pride stand in the way. Pride comes before a fall, as they say.

My daughter now does the training, so it is never concentrated into a set time period, like 4-8 weeks with a conventional breaker. The horse is never pressured. Without time constraints, the horse receives plenty of natural breaks, due to weather, illness, competing, family events etc. The horse is accustomed to being left for days, weeks, or sometimes months before continuing with its breaking.

Lunging

Lunging can be done in a round or a square yard, or in the open. In the early days we lunged in the open, because of lack of facilities. Always use splint boots while lunging and sometimes you might need bell boots if they are overreaching, which is common in young, unbalanced horses. Ask the horse to work around you on a long lead rope at first. As he works out what is required, gradually lengthen the distance, so that he can both walk and trot. The larger the horse, the larger the distance required for it to do so comfortably.

We have rarely used verbal commands until lunging begins, nor have we used much voice in our interaction with our horses, relying almost entirely on the use of body language. We use commands of "walk-on", "trot", "canter", "whoa", in different tones for each command, so they are very distinct. "Good boy" is used after a whoa or walking up to the horse to give it a rub. "Good boy" is about the only word it has learned prior to this. These words are reinforced with body language, and the lunging whip (or you might use a "flag," but a horse can easily become unresponsive to this). Watch liberty trainers at work, they never use flags.

The biggest mistake novices make is to assume that their voice is important. It is rarely consistently used, neither is clicking. We don't click to horses, because buyers use clicking indiscriminately, and the horse has no understanding of it. I often ask visitors, "why did you click?" "What does it mean?" They can rarely explain why they did it. If you use methods which are not universal, the horse has to learn everything anew with each owner. It is a credit to the intelligence of the horse, that it is able to do this at all.

If you don't believe me about body language, stand in the middle of the lunge ring, without turning/moving to follow your horse, and use only voice. Only the occasional horse will get it, and that one is usually way more advanced in its overall training. It has learned to pick up on your intent, through your stance and your eyes.

Asking the horse to stop is very easy by simply heading in front of the horse towards the fence, creating a block as you say "whoa". Teaching the horse to fast stop, is very easy by running faster to the block. "Whoa" is easily transferred later when you ride it. If it tries to go faster when you go to block it, choose a spot further in front of it, and run to that spot, simultaneously blocking with the lunging whip. Each time the horse stops, walk up to it for a pat, or stroke its face across its forehead, then closing its eyes. It is amazing how many horses find it difficult for you to walk straight up to them and give them a pat on the shoulder until you do it a few times. You are waiting for a chew response, usually accompanied by a swish of the tail, a relief response.

The pause is as important as all other parts of your training.

Stopping the horse at the fence, and not allowing him to turn in, is a significant part of its training. We'd like it to stop square and motionless, waiting. The horse only comes into the centre if asked with the "come-forward" feel of the lunging rein. The reason we use a lunge rein, rather than free lunge, is because that is easily transferred into lunging on a showground, or any place where facilities are not available. You can lunge from a lunging cavesson (preferred), a halter, or the bit if the horse is rather strong. As soon as the horse is lunging fluently both ways, he will easily free lunge with stops and turns if required.

You can lunge over a pole, or tarpaulin or tie sacks to the roller. Add in some variation.

In between, rub the horse all over with the whip to be sure he is not afraid. When he understands this, you can progress to lunging with the roller. Attach this loosely at first with a breastplate, (or we clip the sidereins in front) so that it doesn't slide backwards. It can be increasingly tightened as you note the reaction from the horse. You don't want to do it up suddenly to startle the horse into bucking.

The horse which cuts corners is usually dropping its shoulder in or running away from pressure. It is important that you keep the bend with the lunge line, so he doesn't develop a habit of doing this. Flap the line towards him to send him out again, or flick him in the ribs behind the girth with the whip, to establish some bend. You may need to bring him much closer to you to bend him from the nose, which is why a cavesson is used in classical lunging, or you may need to move him further away from you to achieve bend, depending on the horse. Until he can bend to the size of the circle you are on, he is not balanced. He will show his balance by dropping his nose to the ground, sniffing the ground, or stretching down without changing rhythm. If he can do this for a full circle all the better!

The benefits of lunging are enormous: legging up, general exercise, fitness and rehabilitation, assisting balance, development of muscles over the top line, encouraging him to round and reach, and learn voice commands. From your perspective, it is easy to note his potential, his paces, his athleticism, his character, his progress and willingness to learn.

Lunging with the bit, Kyabra Park Simply Red -
owned / photo by Tegan McKenzie

Mouthing

Progress to introduction of the bridle and bit. I like to use clips on the bridle for the bit and reins, so that they can be removed at will for different horses. The mouthing process is best done in conjunction with lunging, and/or used as a break from it. A horse which has never been bridled will always chew on the bit for some time before realizing it is unnecessary. We don't use a noseband.

There should be no space between the bit ring and the corners of the mouth, for the bit to slide. Special bits are not necessary for mouthing, but comfort is. A D-ring snaffle is perfect because it prevents sliding and guides the horse in the direction of the turn by pressure on the opposite side. Others use a lip strap for the same effect. Make sure it is smooth, not old or rusty. If the horse is inclined to get its tongue over the bit, raise the bit up another hole or two, or change the mouthpiece of the bit. If the horse has a very low palate, it may not be happy in a jointed mouthpiece, and a half-moon may be better. I like a thinnish mouthpiece for breaking, as the young horse's mouth is often too small for a thick mouthpiece. However, a horse with a super-sensitive mouth may be far happier with a thicker mouthpiece.

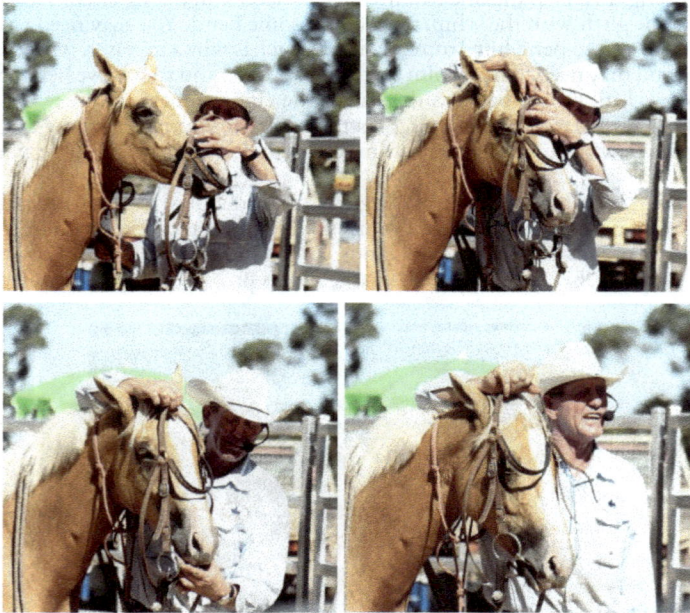

How to put a bridle on correctly- with Col Byron

Photos by Catherine Mahnew

The next step is to hold the reins over the neck in the same manner you do when riding, or "driving over the neck." Hold a dressage whip on the opposite side. If the horse is so tall you can't reach, you can do it under the neck, making sure you hold the reins at the same angle you would if riding. Gently ask the horse to turn its head in response to a gentle pressure and release each time it responds. You want it to stay standing still. Once this occurs you can establish more bend. In between, tap with the whip and ask him to go-forward together with "walk on". To stop, you say "whoa", and stop, waiting for him to stop. Continue in the same manner until he can bend, turn, go-forward, stop, and back up easily by asking with a gentle contact with the hand, from both sides.

This is the best time to introduce stepping away from a touch behind the girth. Use the knob of your whip to ask him to step sideways away from it, replicating your heel or spur. You are asking him to step across with his back legs, both with and without the head bend. Again, do it from both sides.

Only then should you progress to lunging with running reins, or ground driving, or both, depending on preferences. If you use sidereins, make sure the outer rein is much longer, so the horse is still able to bend its neck. With sidereins, or running reins, or a market-harborough, the aim is *never* to pull the head in or down, but to block the horse from holding its head too high or poking its nose too long. By doing this, the horse starts to realise it can take the reins softly and make a steady light contact in its natural frame, without interference from a rider's hands. *We want the horse to find its own sweet spot between the reins.*

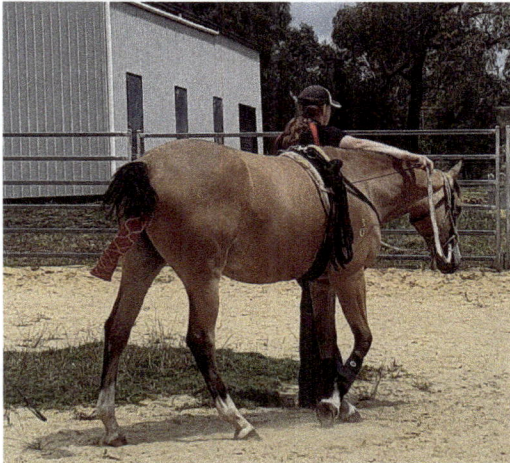

Driving over the neck, Kim Ide with Chalani Topaz

At this stage we will lead the "breaker" off another horse, both with and without the saddle, so he gets used to the height of a rider, and the feel of the saddle with the stirrups dangling. Horses which have been rugged or had a roller on rarely have issues being saddled. He is trotted out of the arena from the other horse to show him some of the sights he will see when first ridden.

The idea is to have him lead with his head close to the rider's knee. The breaker may not understand initially, so you can take a second person with a lunging whip to follow along behind if required, or carry a whip and tap him with it. If he swings out, ride along a fence line. He will soon get the idea.

Make sure the saddle is a good fit. If it is likely to move use a Y breastplate, so it won't roll or slide back. A poorly fitting saddle will create problems unnecessarily. Use a mounting block or stand to lean over the saddle from both sides. You can also do this from a rail in the yard if the horse is tall. Step up and down in the stirrup, holding onto the front of the saddle. Rub him all over and put your leg on his rump. Most horses will move to regain balance at some point. Just keep leaning. Go with it until they stop. Pat him on the opposite side then slide down. Keep repeating this until the horse remains still. Do it from the opposite side. Until he stands still when you mount, don't move to the next stages.

This is the time many trainers will introduce a horse to hobbles and the stock whip but some will only do so if you request it.

Leading a youngster from another horse. Kim Ide at Chalani.

First ride

When all of this is accomplished, it is simply a matter of putting it all together. If you prefer, you can have an assistant at his head, to lead off initially, while you are leaning over him. The next day go over everything, each time taking a little less time, and spending more time in the saddle. At first just ask the head bending exercises, which he knows from the ground, letting him sniff you, in the saddle.

At some point you will feel confident in his manner to ask him to go forward, combining the "walk-on" voice, and tapping with the whip behind the girth whilst applying light touches with your legs. Start with large circles, some one-rein stops from the walk together with voice, gradually increasing the questions. Over the next few days ask for trot, and when this is smooth, a canter. If the horse is not very forward, it is best to move outside the yard to an arena, following another horse. Better still, do a ride-out with another horse. He will pick up the aids quicker. Often the horse's first canter is easier to ask this way. Don't bother with which leg he is taking until he is back in the arena. After a few rides he will pick this up. We like to carry a whip everywhere, as this is a tool he knows from his groundwork; though it is rarely needed, the one time you don't, is the time you wish you had!

Stopping in the early stages should always be the one reined stop, accompanied by "whoa" which he already knows from your groundwork. If he starts to forget, becomes a little strong, or you sense a problem, do this exercise repeatedly until you have some softness and control. Make sure you can do this from both sides. You can ask for bend of the head right to your knee if necessary. Just hold him there until he relaxes and ceases to move.

It is better to do long slow work at first, with natural obstacles and quiet ride-outs, occasionally alternating with bursts of arena work for circles, and further polishing with cavaletti, trail rides or following cattle. Time spent on all the preparatory work means the whole breaking process is seamless. Give him a little feed as you untack him. A tub of water kept where the horse is untacked means that bits can be rinsed before bridles are put away. If at the end you have tail swishing or head tossing, or the horse travels below or above the bit, something is not right and should be investigated. Don't GRIND on your horse. You should have a happy horse at the end of it.

There are many different ways to break in horses, but this is just the way we do it as it is tried and proven to work for all types of horses. When I first started out, due to a lack of facilities, I broke all of my horses in an open paddock and in a dressage saddle, with no issues what-so-ever, simply because of the extensive groundwork given.

Breaking in is the most significant training your horse receives. It adds value to your good horses, but rarely adds $ value to poor types. Your job as a breeder is to produce good types of sound, happy horses, well-trained for their future.

Preparation before the first ride. Breaker (by Nonda Charlie Peno) and lead horse (by Chalani Sunstream) both bred by Bridgette Hasting, rider Lily Johnson. Photo by Priscilla Oberreiner.

How to bridle a head shy horse:

Buckle a strap around the neck. Pass the loop or the reins through the strap from behind and place the loop around the nose. Gently pull down until the loop presses on the soft bones of the nose. Maintain a firm pressure to prevent the horse from lifting its head. Using your free hand put on the bridle. If you can't manage it alone, ask a friend to keep pressure on the loop around the nose. When the bridle is in place, unbuckle the reins at their junction and withdraw them from under the cheek straps. There is no need for excessive force and most horses give no further trouble after they have been bridled a couple of times in this manner. Always check the teeth and poll if problems continue.

Give the young breaker a lot of exposure, lots of ride-outs. Kelsey Stafford on Tarrawonga Phoenix, Leichhardt River, QLD.

Young horse lessons for Chalani Scotch HSH, with Cody Wilson. Photos by Kim Peterson.

Glencoe Sure Thing HSH with 14yo Aiden Elliott, Lissadel Station, photo Kelli Elliott, and top, with Lachlan Glasser. Photo Melinda Fanning.

Chapter 28

Stud routines

Better they say, "I wonder why they gelded that colt?" than "I wonder why they didn't!"

Gelding

The main reason for gelding is for performance. Geldings are easier to handle, more focussed on their work, and can live a happier life with others.

The usual procedure for us is to have all colts gelded and branded just before weaning. The foal then returns to mum for comfort and quickly forgets the event. It also helps avoid swelling because the foal moves around with its dam. If you do it after weaning, you may undo a lot of the trust you have developed in the youngster at weaning time. We do it then because most of our horses will go to their new owners once weaned.

A colt can be gelded at any age, even as older stallions. Some are gelded because they are pinching their testicles or have become too heavy-fronted, which impacts performance. This happens quite often in racing and polo. Heavy necks are frowned upon in the working horse world due to extra pressure on the forelimbs. Geld stallions rather than sell to where they will have limited use and be kept separate most of their life.

It should be the breeder who decides which are permitted to be entire, not the new owner.

Most gelding is done lying down, but some vets prefer a standing procedure. This may be a problem if the colt has a scrotal hernia and some of its bowel descends unexpectedly. If the colt is monorchid (or a "rig"), it is mandatory to take both testicles simultaneously, so abdominal surgery is required to remove the undescended testicle. Never use a rig to sire foals, as the condition is believed to be inherited.

All breeders should take the responsibility to geld seriously. Too many colts are not good quality, which reflects negatively on the breeder and the breed. If you decide one is worth keeping entire, then run it on yourself, and sell it later if it still meets your standards and you have a suitable colt buyer. If in doubt, geld.

We have taken great care to ensure the few colts we have sold are worthy and go to worthy homes. Over 50 years, we have mostly only sold stallions that we have used ourselves first, as proven, genuine sires, and a handful of colts that have since proven our faith in them because their owners have gone on to produce quality stock with them.

"As a breeder, I believe we have a responsibility to ensure the best of homes for our foals into their adult lives - this is why we have a NO COLT policy.
- Peter Woodger.

Beware of the buyer who promises to geld. So often, this will not happen. Make sure you put "gelding" on its papers. I believe that if every breeder gelded everything before it is sold, and advertised a "No-colt" Policy, buyers would need to prove themselves to you and why you should sell them a colt as a future sire, when you will not do so to anyone else.

Unfortunately, this will not stop people from buying colts at sales, as it seems common-place for breeders to enter colts at Breed Sales and let the buyer decide to keep it a colt or otherwise. At some Breed Sales, it appears that at least half the males entered are colts. Breeders need to get their act together and really know that if a colt is not what you'd retain and use yourself, it should not be sold as a colt!

The only exception to this might be a high end performance sale, or the racing industry, where you are already a stud with a strong reputation.

"The colts we produce are of a high quality but as they were sold through the NCHA sale as long yearlings we felt it too soon to geld them, but rather allow the buyer the choice once the colt matures to see how he copes with increasing testosterone and his ability to focus on his training at the same time. One such colt we held in very high regard, but he was not able to focus so they chose to geld him. And out of the 5 colts we bred, only 2 were kept entire and one of those was later gelded too. Because we were selling to a specific industry and educated horse people, the competition was first and foremost in their minds." – Karen Trun, Cutting horse breeder.

There are some breeds where a gelding has little value. Historically, this occurs because too many colts are kept, with only fillies seen to have value. If the breed sees little value in its geldings, and breeders are breeding fillies to breed more fillies, it promotes a never-ending cycle of geldings becoming more worthless. Some breeders send all their colts to local sales just to avoid spending money on gelding and breaking in because they cannot be valued higher than the expense of doing so. This is the epitome of irresponsible breeding, with the wastage noted by welfare advocates to the detriment of all breeders.

If geldings are sold at solid prices, they are marketed on their performance or potential. A good gelding average measures how well the public sees the breed as a whole. That augers well for the future of the breed. Breeders need to raise the bar with their broodmares, keeping those of high enough standard that could potentially produce stud colts. Without that potential, perhaps they are not of a high enough standard to be in one's stud.

Farriery

A good farrier will understand conformation/movement and empathise with horses. He must also have excellent anatomical knowledge of the horse's foot, and be well trained in his craft. He must be able to explain what he is doing so you understand the processes, and you must be able to explain to him any changes, such as in its gait, that you may

A good gelding will always showcase the breed. Waler gelding Wendara Silver Lining owned by Penny Bieber, ridden by Emily Wonka. Photo by Katy Driver.

have noticed since the last shoeing or trim. Farriers spend a small amount of time with the horse, every 5-8 weeks, so it is important to communicate any soreness problems, or changes to its feet. He must also be reliable or he will soon find himself with only those difficult horses or owners which no other farrier wishes to work with.

A good owner needs to have his horses caught at arrival, well handled, and in a spot out of the weather. With larger operations this may not be possible and the horses are done out in the field, possibly in a collection paddock. The owner will book the next appointment, and not cancel at the last minute. The owner will also pay on time!

It makes no difference if you are a "barefoot" fanatic or a "hot-shoe" advocate to the horse; what matters is that the foot is shaped correctly to the horse's conformation, posture and way of going, and he is comfortable on the surfaces you are working on. If he always stands over his front legs, or leans back, has a shortened gait, or a misshapen foot, these all need to be assessed, and a programme of correction commenced.

This starts with the foal at about 2 weeks old, continuing thereafter as required, particularly if it toes in or out, is back on its heels, or shows signs of a club foot. Most will not need anything done with them till they start to outgrow their foal caps at about three to four months of age, where a simple rasp might do. For this reason, you can probably do most of your young horses yourself, for the first year or so; simply buy yourself a good rasp and hoof cutters. (GE brand recommended – expensive but well worth it.) Have your farrier show you how to go about it, that way you can work on youngsters every couple of weeks, rather than wait for the farrier.

The most comon thing judges notice is poor feet, or poor farriery. Neither should be excused. The higher the degree of performance required, the greater the importance of attention to detail to optimise the horse's performance. It is important to have correct boney column alignment and balance the hoof to this alignment, looking both from the front and the side. Farriers must work on correcting caudal failure in shod horses (which often has a conformation base – see p45), while respecting the hoof capsule, for creating and maintaining overall comfort of the horse.

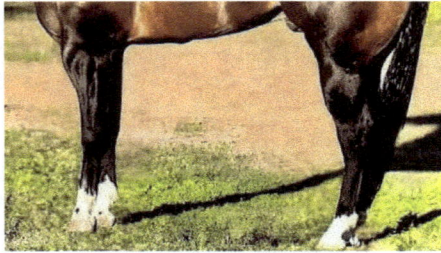

Club feet on both front legs, together with leaning back posture, indicating discomfort in the front legs.

Common issues farriers see are caused by owners not looking at their horse's feet between trims, such as thrush. This is an infection found particularly in wet environments, easily corrected if found early enough, with equal parts apple cider vinegar/water, and sometimes a little copper-sulphate added to help dry out the hoof. Hooves should be regularly picked out, to be sure there are no foreign objects trapped down the side of the frog.

You can make your own hoof grease by rendering down unsalted mutton fat from your butcher, and adding neatsfoot oil half/half, but this will vary according to summer or winter. Apply it to the coronary band and underside of your horses on a regular basis with a stiff paint brush. Always use the natural neatsfoot, not the pretend brands made from chemicals. Keep a pot handy for any time your horses are brought in for handling. In very hard dry conditions, it can sometimes pay to have a "muddy spot" deliberately set up around the water trough so your horses must stand in it. Your horses and farrier will appreciate this.

Dental

It is important to understand that many behaviour problems start out as dental problems. Tom Roberts used to always say whenever you have a training issue, first check the teeth! Horses' teeth wear as they graze due to the abrasive effect of grasses, but horses on hand-feeding regimes may not wear them at the same rate nature designed. The horse's teeth are continually erupting to balance out the wear, but in doing so, as they age, less of the root remains in the gums, and the teeth can become loose or be lost altogether. Most aged horses lose condition because of teeth issues, so you must carefully manage them to avoid this.

Aged horses may have wavey teeth, "rubber" mouth, teeth worn to the gums, root reabsorption, or absent teeth. Young horses may have wolf teeth to remove, sore gums and abscesses as new teeth come through, and caps. Finding a "cap" in the feed tub is not unusual. Adult horses may have fractured teeth, a lost tooth, something caught between the teeth or the side of the mouth, decay, or unusual formations causing ulcers, to name just a few.

These can cause considerable pain, loss of condition, and behavioural issues. Some indicators may be the horse having difficulty chewing feed, being unable to bite into a carrot, dropping feed (quidding), head shyness, or losing performance (e.g., running off the track). There are also changes in head carriage (twisting, tilting the head, snatching, leaning, excess saliva, drooling, yawning). In addition, there can be weird anomalies, like the "ear tooth." The list goes on.

These can develop into more severe debilitation and health issues if left untreated, for example, infection, choke, gastric ulcers, or colic.

Our horse's teeth are routinely checked annually from two years and up, even for horses not in work by a vet specialising in dental work, so that sedation and any other medications or necessary treatments, can be given. They are also weighed at this appointment. Everything is recorded on HorseRecords. We take great pains to ensure their mouths are in the best shape to avoid future complications.

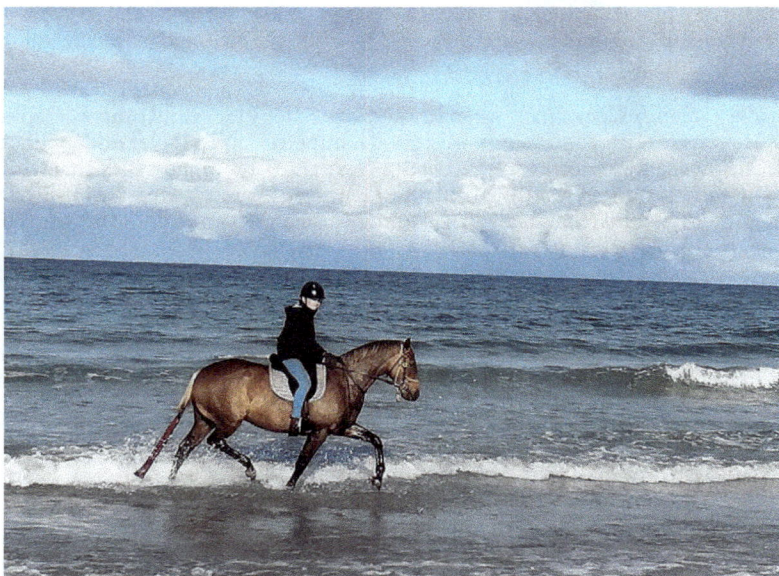

Chalani Sangria HSH with Ashton Ide. Photo Kim Ide.

Branding

In the past, branding was somewhat horrific because the youngster was branded as an adult, brought in unhandled, and often this was his first experience with man. Pictures show great amounts of smoke, wild horses, and tough men doing the job, designed by the photographer to impress the magazine reader.

It is quite different now. Freeze branding is now the method of choice, and brands are much smaller. The hair is clipped and oiled, so the brand doesn't need to be held long.

The horse is branded young, with a simple restraint of twitch or neck hold, in a contained space, such as a stable. Any reaction usually happens after the brand is removed. A good brander knows his job and is very quick. Within a day or so, you can rub the area, and there is no discomfort.

We only microchip if the new owner requires it. It is a simple procedure of clipping a small square at the top of the neck and pressing the microchip gun onto the spot. The number is recorded at the time by the vet and on HorseRecords.

Bodywork

Acupuncture, massage, and chiropractic are important to consider after breaking in and during regular work. This will also benefit serving stallions or any other horse which doesn't act quite "right," such as becoming head shy, "girthy," or a shifting lameness. A lot of horses suffer pelvic and hamstring issues once they are worked. They may be difficult to pick up the back feet. My daughter has massage qualifications, and additionally, we use a person with chiropractor credentials to come in regularly to check our riding horses.

Don't underestimate the value of this form of treatment. We first used body manipulation on our horses 40 years ago, following a pulled-muscle style injury of one of our sprint racing Quarter Horses, which would have resulted in him being scratched from a race. The bodyworker "manipulated him" three days before the race, and he walked off sound, much to our disbelief. He won the race.

Many advocate the usage of massage machines, kinesiology tapes and other therapies, but make sure you are using something with a good reputation, and people with qualifications in their fields. Ice water for legs, is excellent as a swelling preventative, as are astringents and kaolin clay products, available from your equestrian supply store. Swimming too is excellent for fitness and rehabilitation. Consider beach rides as part of the horse's training.

Vaccinations

Apart from tetanus vaccinations, strangles is another vaccination routinely administered. Immunity of foals comes through the colostrum if the mare has been vaccinated two to four weeks before foaling. The foal can start a course from six weeks old when the immunity wanes. A booster is required at the yearling stage and after that, as your vet recommends. Tetanus is essential, because horses as a species are the most prone to contracting tetanus, and it is nearly always fatal.

Some vaccinations are "risk based," administered in certain populations after assessing risk vs. benefit, depending on regional disease variation, competition or transport requirements. Strangles is one of these. Strangles requires three vaccinations, two weeks to a month apart, and an annual booster and these can be given together as a "two-in-one with tetanus." Another is Hendra.

Hendra virus is a bat-borne virus that is highly fatal in horses and humans. Numerous outbreaks have occurred in Australia in Queensland and NSW since it was discovered in 1994. People can become infected through contact with an infected horse. After incubation of 9-16 days, severe respiratory illness and flu-like symptoms progress in unvaccinated horses, with over 50% of cases resulting in death. Vaccination is the only prevention where flying foxes (bats) are common. Hendra vaccination is mandatory to compete and travel horses in some areas. For complete information on the virus, and vaccination strategy, go to the Hendra Virus Information Pack   https://www.daf.qld.gov.au/hendra

Worming

Most studs will give a paste wormer and rotate as their vet advises. It is generally recommended to do a fecal egg count at least once a year, taking samples from each paddock or suspect horses. Some studs are set up to do egg counts themselves. This is to determine which horses have heavier worm burdens and which are prone to "shedding."

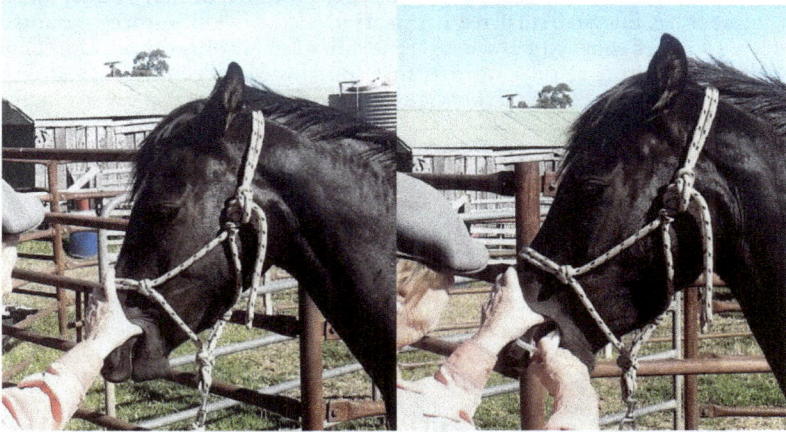

How to worm a horse correctly with thumb in mouth.

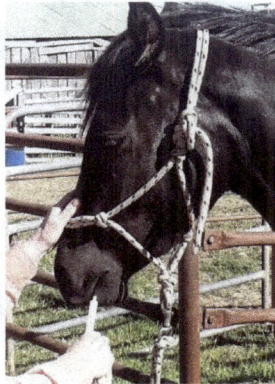

Incorrect

A good count is between 0-200 EPG (eggs per gram), and a moderate one is between 201 to 500. If you continually get higher than normal counts, you may already have a resistance problem in your herd, or the individual has a lower immunity than others. You must adopt a targeted programme, and follow-up worm egg counts with these horses.

Another approach is to do a count before worming, then two weeks afterward. The count should have dropped by about 80%. Check also that you are not under worming or that your horse is spitting it out. If you suspect the latter, try putting a wormer in its feed. Most of these are much more palatable than they used to be. Finally, check on the label what your wormer is effective against and how long it is effective. For example, some may not be effective against bots or tapeworms.

Small redworms and large strongyles (bloodworms) are the most common in adults. The larvae can "encyst" and quickly burrow into the gut lining. They can cause severe colic by causing blockages or destroying arterial walls, blood vessels, organs, and tissues, often with fatal consequences. Redworms can appear as tiny pieces of thread in your horse's droppings. They are very short, thin, and can be hard to see. They are white if they haven't yet fed and red if they have. Weight loss, diarrhea, or colic may be early indicators; don't ignore these. Encysted small redworms are *not detectable* in worm egg counts. Only two wormers target encysted redworms: Fenbendazole and Moxidectin. Don't confuse this with Ivermectin. The only wormer that contains Moxidectin is Equest. Five-day PanacurGuard contains Fenbendazole. Equest is also the only one that doesn't harm dung beetles.

Watch any chemicals you use on your pastures which can affect bees and dung beetles – definitely nature's friends.

Young horses in the first year are most susceptible to ascarids, or large roundworms, which can quickly develop into large numbers and cause severe debilitation or even death. Signs are lethargy, pot belly, loss of weight, and diarrhea. Time one of your wormings a month after bot fly season, so you can treat any lavae in the stomach, before they cause ulcers.

Fannie Bay Equestrian Club, Darwin. Retired and aged horses still giving pleasure. Sea water is excellent for skin complaints.

Other parasites

Ticks are best treated by spraying with a pyrethrum-based insect repellent. Wait for about an hour, and then apply the solution again. The ticks should fall off, or you can remove them with tweezers, as they should be dead.

Queensland itch (or sweet itch) is an allergic reaction to the saliva in midges' bites, creating intensely itchy lumps. In chronic cases, the skin becomes thickened and corrugated. It is a common occurrence in Queensland and northern NSW in summer. A weeklong daily wash with pyrethroid is recommended, then once weekly until the condition is under control. Cover the horse with a sheet or shade cloth rug, with a neck rug attached for best results. Some horses will get this every year, which is extremely hard to manage. Unless this horse is important to you, it may be better to sell it down south. Don't forget such sensitivity is likely inherited.

Ivermectin wormers have been quite effective in treating external parasites, such as lice, ticks, and skin-dwelling larvae of parasites such as Onchocerca and Habronema. It is relatively safe in horses. If you see external skin parasites, it is good to give them a medicated shampoo wash once a week for two weeks, keeping the shampoo on for approximately 10 minutes before washing it off. Then rinse thoroughly.

Where this cannot be managed, you can use a lice powder. Quarantine any horse affected with lice. For any other non-specific or non-identified itches we have used Calafea Horse Oil with good results.

Infectious conditions

Fungal skin conditions, including ringworm, spread on tack such as bridles, girth, and saddle cloths, during transportation or being housed at a showground. The gear will need to be treated, as well as the horses. These infections are highly contagious. Humans can carry viruses on the skin, clothing, and shoes and transfer them to other horses. Whole groups of horses may become infected. Your vet can take a culture to determine the type and commence treatment. Horses with contagious conditions are prohibited from travelling or competing, but it is often not realized they have the condition until approximately three weeks after infection. Anti-fungal washes, often several different ones, will be necessary before you find a successful one. Gently lift scabs with a scouring pad so the treatment can reach under the scabs. Apply the treatment according to the directions.

Juvenile warts are a viral infection likely to occur weeks before any horse shows signs. It nearly always affects young horses or those not exposed to the disease before. The warts tend to grow all over the muzzle and, sometimes, in extreme cases, around the eyes. They can be quite painful. Quarantine the affected horse. Treatment is limited. Eventually, they will fall off spontaneously after a few months and the horse will not get it again. He may need extra feeding if it becomes hard for him to graze. Once it is on your stud, you likely will get repeated outbreaks with each new crop of youngsters until it disappears altogether.

Colds are infectious, but mostly in younger horses because older horses have acquired immunity. Bacteria or viruses cause colds. The foal can be quite lethargic and have a runny nose or thick discharge, and he may have a cough. Check if he has a temperature and if so, speak to your vet. Do not work a horse with a cold. Isolate him, so he doesn't spread droplets to other horses.

With all the different infectious diseases in Australia, we have very few compared with other countries. Our strict quarantine laws have made new diseases difficult to bring into the country. In August 2007, the EI virus came to Australia due to the failure of bio-security measures. It infected 69,000 horses, and 9,600 premises were in lockdown from horse movement. It was a huge financial burden on horse owners' livelihoods, the agriculture sector and taxpayers, but the disease was eventually eradicated.

If you suspect any infectious disease, isolate the horse immediately. Any new horse brought onto your property, a horse bought, or a mare to your stallion should be quarantined away from other horses for 14 days unless you know exactly where it has come from. Do not allow nose-to-nose contact with other horses during this time. Take extra precautions if the horse has come from an area known to have an outbreak, such as Hendra or strangles. Check on the vaccination status and make sure your horses are also vaccinated.

Thoughts on Potentially Overwhelming the Immune System

"Erring on the side of caution, I never combine multiple vaccines, sedation or dewormers - nor do I administer successively within a short period of time. Why? Combining or administering successively compromises your horse's overall well being. It also introduces a wide array of challenges to the immune system, which weakens the ability to respond effectively. The immune system is like an army. If half of the soldiers are battling adverse effects or appropriately reacting to a vaccine, the capacity to respond to additional threats is reduced.

"<u>Vaccines:</u> If you combine and your horse has an adverse reaction, you may not be able to attribute it to the vaccine - or determine if you should use that brand again. In addition, the vaccine's effectiveness may be compromised if the horse's immune system is not able to respond appropriately.

"<u>Sedation/Tranquillizers:</u> Side effects can include but are not limited to: salivation, sweating, diarrhea, elevated blood glucose, decreased intestinal motility, labored breathing and reduced heart rate. These cause additional battles for the immune system to fight.

"<u>De-worming:</u> Together, the chemical, combined with parasite die-off, releasing parasitic endo-toxins, can account for quite a toxic load. Often times individuals with a large parasite burden are already compromised and may be subject to impaction colic due to the sheer volume of parasites to excrete."

– From the Hay Pillow Blog (highly recommended resource for care and nutrition)
https://www.thehaypillow.com/blogs/news

Thoroughbred gelding Strategic Glass, tent-pegging in style. Owned/ridden by Tim Goodes. Photo Teena Goodes.

Chapter 29

First aid

Dealing with injuries and disease

First aid can be complicated or simple, depending on how much access you have to your vet and how much you can do yourself. Old broodmares that never leave the place may have scars for life without it mattering greatly. Semi-serious cuts may only need hosing down, an antibiotic spray or Flint's oil, and fly spray, until healed. Wounds that need stitching will need a timely visit from the vet, or the skin flap may dry out too quickly to be salvaged. Most will need bandaging, especially if the wound is on a joint, and if there is joint fluid leakage, antibiotics. You will probably find you can remove stitches yourself with sharp scissors.

Do you know how to bandage a horse properly? If not, take a course or have your vet show you. Having the vet out whenever a bandage needs changing is way too expensive. The same goes for a course of injections. Your vet will advise you on how to give an intra-muscular injection.

Bandaging will be essential for young stock, show stock likely to scar, or deep wounds. Ensure tetanus vaccinations are up to date. Depending on the severity, you can generally leave a stable bandage on for four to five days. If there is a lot of oozing or pus, it has to be changed more often. Allow for swelling with some vertical snips of the bandage at the top and bottom, as you don't want it to act as a tourniquet. When in doubt about any procedure, speak to your vet.

Bandage scissors

Have plenty of elastic bandages, cotton wool sheets, and gauze. We have managed to obtain some past their use-by dates from a contact in a hospital. Swimming pools often have towels in their lost property they are glad to donate. Think outside the box on how you can obtain useful items.

Have a supply of syringes of different sizes, needles, sterile blades (for cutting bandages and proud flesh), vaccines, and a fridge for storage. Bar fridges are cheap these days. A little copper sulphate crystal, mixed with topical ointment, is excellent for applying to proud flesh. However, the wound must heal first; otherwise, you will delay healing. You may have other stud supplies, such as progesterone, oxytocin, antibiotics, and pastes, by arrangement with your vet. Always have a thermometer in your first aid kit. Additional items include antiseptic creams, purple spray, zinc cream for noses, greasy heel treatments, fly spray, scalpel blades, scissors, disposable gloves, soap, and clean buckets.

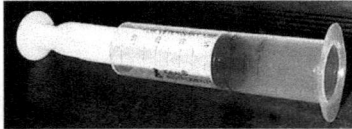

Syringe to use as an udder pump.

Have a couple of spare twitches, one for adults and another for youngsters. You can also buy a metal nose twitch that doesn't need a handler to hold. This is quite effective on some horses, allowing you to treat them without an extra person. I like having a waste bin right next to the stable and vet area.

Vital signs of the normal horse are:

- Temperature: 37.2 to 38.5 degrees centigrade (99.0F to 101.4F).

- Pulse: 28-44 beats per minute.

- Respiration: 10-24 breaths per minute.

- Mucous membranes: (gums): Moist, healthy pink color.

- Capillary refill: Two seconds or less.

Label this on your first aid kit for your family or staff, and provide a list of emergency contacts. It pays to have human aids in the kit, too, such as band-aids, tweezers, sunscreen, etc. Keep a spare emergency kit in your float or car. A fishing tackle or tool box makes a good first aid container.

Only use eye ointments when the vet prescribes them because they expire quickly and may not be suitable. The wrong one may do nothing and instead delay healing. Eye treatments are best left to the experts.

Choke means a horse is unable to swallow its food properly. It is usually first noticed as a heavy green discharge from the horse's nostrils and a refusal to eat. The horse may be coughing. Massage the neck of the horse to find the ball of food and see if you can dislodge it by stroking your hand firmly down its neck. You can hose its mouth to see if you can add some lubrication, but do not force water down its throat; you don't want water to enter the lungs by mistake. If the horse does not respond to your efforts, it's time to call the vet.

If you strike a case of colic, treat this seriously. If your horse is restless, getting up and down, or rolling, walk him until the vet arrives. If he is calm, check his temperature and gums. Pale gums can be a sign of shock. Pulse or heart rate is a good indicator of pain and its severity. You can palpate a normal pulse under his jaw near the jowl or at the back of the pastern. A horse in pain has a heart rate of more than 60 beats per minute (normal is

48 or lower.) You can take the horse on a float ride to see if it can do manure. But other than that, it is best to call a vet before symptoms become more severe. This is a time when you and the vet will be most pleased if it turns out to be a false alarm.

Other injuries, like a kick to a stallion's penis, should also be taken very seriously. Without treatment, colitis can develop, which is stress-induced and often fatal. In addition, the stallion may become infertile during this time from inflammation and high temperature.

With climate change and variable weather, we are seeing more foot abscesses and problems, such as greasy heel. An abscess grows due to the hoof wall's expansion (dampness) and contraction (drying), eventually breaking down the protective hoof barrier allowing microbes inside. It can also enter via seedy toe, where the hoof wall separates from the sole. Abscesses can be very painful, often causing the owner to think the horse has broken its leg. Minor abscesses will break out at the coronary band after a few days, and the horse will return to normal. Chronic abscesses and horses in work will need the vet/farrier to drain them and give antibiotics and painkillers. You may need to wrap the hoof in a poultice available from most veterinary supply stores, and contain the horse in a yard or stable.

Choose a kit with a handle, and
preferably a "cooler" type bag.

Greasy heel (or mud fever) causes cracks, sometimes bleeding, and scabs or scarring behind the coronet and pastern, especially in horses with white markings. It is influenced by ultraviolet light. It can be difficult to treat, and infection can set in quite quickly, causing the leg to become puffy. Horses can have the infection circulate throughout their bodies, sometimes resulting in death. We have found the often prescribed green lotions of limited use and have had more success with Filtabac, a sun filter and anti-bacterial cream. It is also excellent for sunburnt noses. In severe cases, bandage the affected area and treat with antibiotics. Bag Balm is *the* best preventative for susceptible horses. Owners swear by this product.

It would be best to list emergency contacts, safe sites for fire or other disasters, your disaster plan, and feeding schedules in your saddle room, feed shed, and near the phone where staff or other family members have access. They can be on a whiteboard or noticeboard. It is easy to print a list of each off HorseRecords. I can't stress the importance of good records. You may not need it now, but you will be grateful for the forethought if something goes wrong during a vacation or another time away.

Every horse owner should own a horse float in good order to take their horse to the vet clinic in an emergency. (I don't like selling to people who don't own a float.) In some instances it can be used as a crush. For the breeder, buying a crush should be high on the list, after a tractor. Unfortunately, when breeding horses, you will become accustomed to administering first aid in collaboration with your vet, so establish a good rapport. That includes paying on time!

Your disaster management plan

Think about the possibility for natural disasters and emergencies well in advance.

> *"The property, paddocks and surrounds have been mowed to an inch of their lives, all fire hoses out with irrigators and sprinklers set up and all areas drenched last night and this morning, leaf blowers full of petrol and ready, radios have batteries ready along with all phones and tablets charged to hear the scanners, water trailer full and ready to be put on our spare car, all horses with rugs and masks off and halters on all gates, all agisters are notified and have our fire plan on a whiteboard here ready if they need to come up to help, including knowing what to wear and what to do if they come up to the property, fridges full of cold water with the house closed up to stay cool and safe with all pets inside......and now we wait.....and cross all fingers and toes this will just be another fire drill for the next 3 days!!!"* – Glenferrie stud.

Young horses need a job. Chalani Chiffon ASH mustering. Photo Kelsey Stafford.

Chapter 30

Weight management

Nutrition simplified

In the past, the vet was mainly used for injury and disease management. Disease prevention and healing focussed primarily on nutrition. While not as much was known about nutrition, it certainly was understood by experienced horsemen that a balanced diet was a key to longevity and health. People fed their horses simply, with few difficulties and little need for supplements. Nowadays, with all the marketing hype, owners often feel guilty if they don't feed their horses the latest magic formula. There is absolutely no need to do so.

If buying in hay, ask if it has been tested. First, have your local agronomy department check your soil to see which minerals require balancing. Hay/pasture testing is more accurate than soil testing, if you already have a place. Plant pastures that are considered most suitable for horses in your area. Horses don't need high sugar content in their diets, and in fact, this can be quite damaging for horses' teeth and breeds prone to laminitis or insulin resistance. There is much more dental decay where cereal hays are a horse's main fibre source.

Each feed type has its drawbacks or benefits. Small amounts of clover do not hurt; lucerne (alfalfa) is a great way to add calcium. Barley, and rye grasses have the additional problem of grass seeds and awns getting into eyes. Wheat can cause colic or scours. Bran/pollard should not be fed in areas with little calcium or an oversupply of phosphorus. The calcium/phosphorus ratio should be 1:1 to 4:1. This balance is important for bone strength, cell function, and muscle health. Calcium and phosphorus comprise about 70% of the body's mineral content. In the broodmare, milk production increases calcium demand. Overfeeding phosphorus can cause big-head disease, a typical problem in tropical/semi-tropical areas.

Legumes like lucerne hay are a good source of quality protein, as are lupins or faba beans, which are good for laminitic horses. Ideally choose lucerne hay third cut as less stalk means more protein. Horses without good quality pasture are likely to be deficient in protein. Horses with insulin dysregulation or EMS (equine metabolic syndrome), may not tolerate the high crude protein in lucerne, so it doesn't suit all horses.

Historically, hay was chopped into chaff, or "chop," to enable draft horses to eat out of nosebags. It is not economical to feed chaff to horses, and Australia must be the last place in the world to feed it regularly. The best chaff cutter is the horse's teeth. Your horse's teeth will be better off, and so will the gut! The best way to feed your horse is to have his teeth in good order through regular dental work. Chaff is best for older horses having trouble with their teeth, or young foals.

We don't feed whole oats because, too often, you will find seeds in the manure. (A friend considers this an advantage, because the cockatoos will break up manure in search of it.) Crushed oats are good if you feed them fresh. Otherwise, the nutritional value deteriorates quickly.

Grass (Meadow) hay and oaten/wheat hay are good choices for most horses, and Teff hay for fat ponies. Allowing free choice hay to all but the most voracious eaters is the ideal way to feed. It satisfies the horse's need to eat almost 24 hours daily and discourages boredom. The horse's gut is designed to eat fibrous foods (bulk) as its primary food source. The horse's saliva has limited digestibility. However, it acts as a buffer to the hydrochloric acid in the stomach. Therefore, limiting a horse's chewing time will likely lead to gastric ulcers. If you suspect ulcers, you need to check this out with your vet by laparoscopy as different types require different treatments. We have found Boost (Optigut) very helpful.

If a horse is prone to bolting down the food, put a brick in its feeder, reducing the risk of "choke." Horses are unable to reflux any stomach contents, so once food has entered the stomach, it must pass on down the line.

The horse is designed to be a "trickle" feeder. Its stomach is small for its size and will empty within 30 minutes if it is eating continuously. It may take up to 24 hours if it is fasting. Relatively little digestion occurs in the stomach. Hydrochloric acid and pepsin are released, which begins the process of protein digestion. If the stomach is allowed to empty, which occurs through modern feeding practices and transportation, the non-protected stomach cells will likely ulcerate. Up to 80% of horses not on pasture have some ulceration. This may influence his appetite, behaviour, or performance and may even promote windsucking.

The stomach makes up only 10% of the digestive system. The phrase "little and often" is particularly pertinent to horses.

Most digestion occurs in the small and large intestines. The small intestine and the pancreas release enzymes which break down proteins, fats, starches, and sugars. It is important to feed good quality protein. Horses, unlike the cow, are prone to colic because toxic material enters the intestine and is absorbed into the bloodstream before it can be detoxed. Therefore, it is important not to feed poor quality or mouldy feeds.

Bile constantly flows into the small intestine from the liver because the horse does not have a gall bladder to store it. Bile breaks down fats that can be absorbed and aids in the removal of toxins and metabolic waste.

Once food matter enters the hindgut, the bulk of digestion occurs through microbes producing fermentation. The caecum is the equivalent of the human appendix, but it is huge. It has a capacity of around 28-36 litres (think jerrycan, but larger). It can enlarge or shrink according to the horse's diet and how much fibre it is eating. This is why a horse may gain a "hay belly" or "grass belly" when out at pasture and lose it when stabled on hard feed. Do not be fooled into thinking because he has a belly he must be fat. He may well be but use his topline as the indicator. The caecum can cause problems if the horse eats a lot of dry feed without adequate water, setting it up for colic.

The least labour-intensive way to feed hay is to use round bales in a ring, allowing the broodmares and youngstock free choice. They are hard to dispense without using a tractor, but there are some devices you can buy that tow behind a vehicle. For many years we used harvester tyres with great success, by putting several square bales in them at a time.

Since it can take two to three weeks for the microbes to adjust, only change feeds gradually. Gut microbes will produce all the vitamins a horse requires if it gets a balanced diet which includes *fresh green grass*. If not, salt, omega 3 and Vitamin E are the essentials to supplement especially for broodmares, stallions and horses in work. The latter two are available in unsaturated fat sources such as flaxseed/linseed, wheat germ, or rice bran oils, which also provide extra calories to increase weight. Add this in slowly, up to around a cup per day.

Microbes also assist temperature regulation through the fermentation process, so feeding hay in winter is a good way to keep them warm. Horses find it easier to maintain temperature in cool weather rather than cool off in hot weather. Fortunately, they can sweat, but ensure you provide plenty of salt in the form of a salt lick and water, even spelling horses. Check water troughs regularly for cleanliness, algae, dead animals (birds), or even ice. If water is very cold, they dislike drinking and can dehydrate. Goldfish are a good way to keep troughs clean.

Most nutrients produced in the caecum are absorbed in the large intestine. It consists of pouches, facilitating the digestion of large quantities of fibre. Unfortunately, the pouches can easily become twisted or filled with gas due to the fermentation. The last passage is the small intestine which leads to the rectum and anus. Its function is to absorb liquid and return it to the body. The remainder, undigested material, is then passed as manure.

The process of digestion will take 36-78 hours. Stress, such as weaning, antibiotic use, travel, illness, and injury, can contribute to digestive upsets. Trying to mimic a horse's natural grazing pattern is the best way to reduce problems associated with the horse's unique digestive system.

It is important to treat your horses as individuals and manage their dietary requirements accordingly. Group your horses according to their needs. Young horses go through growth spurts which can suddenly cause them to become skinny if their diet is not adjusted. For example, if one is on the same diet, eats all his food, and is now losing weight, you need to increase his food. Feed so that he starts to leave a little of it. You will then know that you are feeding enough.

Harvester tyre hay feeder

A creep feeder is a great idea for studs to supplement the foal feed. The creep allows the foal to enter but not the mare. It is usually a yard with a top rail a little over the height of the foal, which can then be increased as the foal ages. Creep feeders allow foals to become accustomed to eating solid feed before weaning time, reducing stress. If you feed the foal alongside his dam, there are small feeders on the market which have evenly spaced bars over the opening that prevent the mare's larger muzzle from reaching in to snack on the creep feed. Be careful not to overfeed the foal, as this may affect its joints.

Aged horses, particularly stallions and broodmares, require extra feeding. It may be the difference between getting one last foal or not. You may notice their teeth are compromised or even absent. Most aged horses can be kept in good condition by adjusting their feeding regime to "soft" feeds (such as mashes) and if they can eat it, chaff, as they are unlikely to manage hay. Feed as much as they will eat, and often. Add oils for extra calories. They may be able to grasp long fresh grass with their lips. Assist older horses with a rug. There is no sense in allowing an aged horse to go downhill by not feeding it. It should be well looked after, rugged, and allowed a peaceful retirement or euthanized.

There is nothing wrong with prepared and processed feeds; the feeds must meet quality standards and be properly balanced. It certainly reduces mixing time, and they can be mashed. Cubes which can be be picked off the ground when dropped, provide les wastage than pellets. Many owners do not understand the basic groups, which can be divided into hard feed (concentrate), complete feed, or a combination of both. Concentrates are fed as a *top-up* feed with hay or pasture, where the protein content is insufficient to meet needs due to seasonal changes or inadequate quality.

A complete feed, however, is designed for the horse with little or no natural feed. Horse paddocks easily become horse "sick" with no natural goodness left. The complete feed already has bulk added, which makes it more expensive than giving a top-up hard feed with hay. Owners mistakenly give the complete feed as their top-up, when a concentrate is required instead.

Complete feeds are often suitable for ponies whose diets must be restricted. Ponies on slow feeding regimes may benefit from Fibre Beet, a product that allows them to stay topped up and feel full without ingesting a lot of sugars. It must have water added and be allowed to swell before being fed. I am not a fan of using slow feeding nets with round bales, because of the increased risk of wear to the front teeth.

Always check the label on processed food to determine its use. The highest protein content will be broodmare/stud stock concentrates. This can be fed to growing youngsters. Lower protein foods are for horses in work, and complete feeds generally have the lowest protein levels because of the amount of fibre in them. You do not need to add a mineral or vitamin supplement when feeding these feeds appropriately.

Warning: never feed horses cattle, alpaca or chook pellets, as sometimes they have the ingredient Monensin (Rubensin) as a "fattening agent." Dust can also cross-contaminate into your horse feeds. This is highly toxic over time to horses, humans and dogs (which might eat it) indicated by lethargy, loss of condition, muscle wastage and neurological symptoms. Try to choose brands which only manufacture horse products.

It is always good to talk with a nutritionist for specific needs or to discuss a complete feeding plan. They may recommend blood testing, or even hair mineral analysis, if blood tests do not identify anything unusual. You can balance your horse's diet by using the FeedXL nutrition calculator: https://feedxl.com/australia/

You may notice a horse scouring upon a change of diet. It is usually seasonal because of higher levels of starches and sugars. If this occurs for lengthy periods, take the horse off its regime, and change it for something of lower content. Pre-biotics can be very useful, as can Greek Yogurt, given with a syringe. Pro-biotics are beneficial bacteria, and pre-biotics is food for these bacteria. If scouring continues, do a faecal egg count for parasites, or have your vet tube drench with healthy manure. We have also used Boost with success.

Eating weeds indicates that the horse is not getting enough of the right foods. This can lead to toxicity, especially liver damage. It can also lead to neurological conditions, such as false stringhalt, where the horse develops a "snatching" action with its hind legs. Take the horse off the offending paddock immediately and consult with your vet.

Horses on bare ground may ingest sand, which builds up in the stomach. Avoid feeding directly on the ground and use bins or hay nets. Sand is a leading cause of scours, colic, and unexplained weight loss if not treated early enough. You may have an arrival which fails to do well in spite of extra feeding. Several paraffin drenches by your vet may be required to alleviate this condition. We have found feeding Equilax to be a very good treatment. Others regularly feed mashed pumpkin or psyllium husk powder to prevent recurrence, though there is little evidence of the effectiveness of these unless fed in combination. Studies by the University of Florida, found that feeding hay at 2.5% of body weight daily (11.35kgs for a 450kg horse, or 25 lbs. for a 1000 lb. horse) was overwhelmingly the most effective preventative to move sand from the digestive tract and prevent accumulation.

Your horse's diet will be reflected in its overall health, and its hooves. Hoof rings may form upon change of diet, and poor diets lead to shelly or weak hooves. *The biggest improvement you can make to hooves and health, is to ensure good nutrition.*

If you would like an A4 sized copy of the following chart, feel free to email me with CHART in the heading.

Clydesdale yearlings. Photo courtesy Ian Stewart-Koster

Body Condition Chart

0. Very Poor

- Very sunken rump
- Deep cavity under tail
- Skin tight over bones
- Very prominent backbone and pelvis
- Marked ewe neck

1. Poor

- Very sunken rump
- Cavity under tail
- Ribs easily visible
- Prominent backbone and croup
- Ewe neck - narrow and slack

2. Moderate

- Flat rump either side of backbone
- Ribs just visible
- Narrow but firm neck
- Backbone well covered

3. Good

- Rounded rump
- Ribs just covered but easily felt
- No crest, firm neck

4. Fat

- Rump well rounded
- Gutter along back
- Ribs and pelvis hard to feel
- Slight crest

5. Very fat

- Very bulging rump
- Deep gutter along back
- Ribs buried
- Marked crest
- Fold and lumps of fat

Mares: moderate to good condition. Stallions: slightly fat at the start of season. Paddock horses and yearlings: good with slight changes acceptable according to the season, otherwise provide supplementary feed. Fat horses: monitor closely to ensure weight loss, limit disease.

Chapter 31

Equipment

Equipment is personal preference. You don't need much, especially if you only breed a foal or two. If you go deeper into understanding the horse, if you can learn to read the horse better, you add better techniques rather than a lot of tools or equipment.

Old furniture can make excellent storage. This one has extra backing to hold the saddle rack.

Have covers for all your good gear. Bridle bags, saddle covers and the like, prove inexpensive for the time and maintenance saved. You will need several saddles for different sized horses, so use saddle racks, a sawhorse, or drums on their side for the short term. Hooks in your storeroom should be easily reachable, and best if you can sort everything by size, label, or colour code. I like wheelie bins because I can put rugs in them for storage over winter. I have the occasional one out next to a gate to store pellets, a bucket, and a halter, without the rain getting in. Try to keep spare brushes, bits, clips and rug straps in see-through plastic containers for easy access. A fishing tackle box is excellent for this. Empty feed containers are handy for storing whips. Old bookshelves and drawers make good storage for smaller items. Everything should be lockable and vermin proof. Be creative.

Don't buy cheap gear. It must be durable and robust. You can't risk your tack breaking at the wrong time. Better to buy good second hand, if you can't buy new!

A short-list of items to get you started.

Branding iron	Horse Float and floating boots	Rugs – winter, summer, shade cloth
Bridles and selection of bits	Items for obstacle training	Running reins
Broom	Knife	Saddles of a variety of fit
Clippers	Lead ropes of various lengths	Scissors
Competition gear	Lunging gear	Shin boots
Creep feed set up	Manure pit with tractor access	Show halters
Crush	Neck collars (different sizes)	Stock whip
Electric unit / tape pegs	Neck straps	Tie up area
Farrier tools	Nosebag	Torch
Feed bins, tubs and feeders	Portable yards	Tractor
First aid kit	Pooper scooper	Tyres
Fly veils	Ride-on mower with trailer	Twine
Foaling alarm	Roller	Twitch
Fork	Rope halters for daily use	Wheelie bin
Fridge	Rope 3m/10ft with ring, variety of lengths	Wheelbarrow or skip
Grazing hobbles	Round bale rings	Whip / poly pipe
Hard hat	Riding boots / spurs	Wire cutters
Hay nets/ hay bag		

Horse floats

A float is a major investment, and you and your horse's safety is at stake. If buying secondhand, take your time, do your homework, and be aware of scammers. Have it thoroughly checked by someone experienced. Especially inspect the floor and potential rust.

If you service your float regularly and keep it well-maintained inside and out, it will hold its value if you ever need to sell it. After 30 years, I sold my first float for considerably more than I paid. If possible, keep it under cover, and check your tyre pressure and lights after it has been parked for a few weeks. A two-horse float at the time of writing, can cost upwards of $13,000 and have a tare weight of around 800 kg with an ATM of approx. 2,000 kg. The three-horse float can cost over $23,000 and have a tare weight of around 1,300 kg and an ATM of approx. 3,500 kg, making them, on paper, just towable with many popular 4 x 4 light vehicles.

Your purchasing checklist

- Comfortable space for the horses you are towing. Look at roof height, allowing your horse to stand comfortably

- Electric brakes (in working order) and breakaway unit

- Light truck tires

- Solid, sound flooring (most important!); steel followed by hardwood, layered with plywood

- Strong welding (Australian steel for our terrain) and good construction techniques

- Fully rated jockey wheel

- Good ventilation; sliding windows and pop-up vents

- Removable dividers and bars

- Adjustable chest bar (needed for different horse heights or foals)

- Smooth edges on all outer surfaces; mudguards, tailgates, and access doors

- Safely located recessed tie-up points

- Rear latches that a horse can't get hooked on

- Swing out breech gates that open wide enough

- Fixed rubber matting on the floor, side ramp, and tailgate ramp

- Guard bars over all windows

- Multiple internal tie-up points, recessed near head position.

- Padding on chest bar, centre divider, and all horse surfaces

- Filled in A-frame towbar

- Gas struts on all access doors and tack doors

- Recessed latch on front door

- Covered light fixtures

- Rear upper door or tarp for dust or rain that can be kept open while traveling in hot weather

- Rear holding bracket for superior airflow

- Checker plate protection

From Discovery Horse Floats   https://www.discoveryfloats.com.au/blog/

Towing your float legally and safely

Towing a horse float has the same legal responsibilities as towing a caravan, boat, or large trailer. The driver is responsible for ensuring that the tow vehicle and horse float are loaded correctly, and weights, hitches, and chains are all within legal limits.

The tow vehicle *must* have the capacity to tow the float within the manufacturer's Aggerate Trailer Mass (ATM), the tow vehicle and trailer's tow ball weight (TBW), as well as the towing vehicle's Gross Combined Mass (GCM).

To find out if you are driving legally, the easy way is to weigh your fully loaded vehicle and float on your next outing. Many local councils will allow you to use their weighbridge at the dump, and commercial weighbridges are also available.

However, the mobile weigh companies will weigh each wheel and axle individually, the tow ball down-load weight, and the weight of your tow vehicle. This will give you much more information to see where the weight is. Anything else involves guesswork.

Know your trailer and tow-vehicle limitations. Every vehicle has a maximum weight. The manufacturer sets these and, under Australian Design Rules, legally *must not be exceeded to remain roadworthy and insurance compliant.*

The tow vehicle will have a rated gross vehicle mass (GVM), the maximum weight allowed on that vehicle's wheels. It includes all drivers, passengers, vehicle accessories like bull bar, roof racks, the load, fridges, canopy, drawers, and the tow ball weight of the fully loaded trailer or float once attached.

The tow ball weight (TBW) is the maximum allowed stamped on that vehicle, the tow bar, tow hitch, or tow ball. Whichever is lesser rated of them all is the max weight allowed. All legal 50 mm tow balls must have a stamped maximum weight. Unfortunately, many out there don't comply. The 50 mm ball is limited to towing a trailer/float up to 3,500 kg.

The tow vehicle's towing capacity must not be exceeded and varies between many 2 x 4 and 4 x 4 models and makes. Just because the vehicle is marketed as having a 3,500 kg tow capacity, this measure is often restricted to just a driver with no passengers or load. This rarely happens in real life, so do your calculations.

On the average tow vehicle, a load of 200 kg on the tow vehicles' tow ball will equate to a load of around 300 kg on the rear axle due to the leverage effect as it is taking weight off the front axle. (A multiplier of 140 to 150 % of the tow ball weight). The tow vehicle will also have a maximum gross combined mass (GCM). This is where many vehicles become illegal.

The trailer also has a gross trailer mass (GTM) which is the maximum weight that the trailer can weigh on the wheels. It excludes the actual tow ball weight, which becomes a load weight on the tow vehicle when the two are coupled.

Many tow vehicles have manufacturer-recommended towing speeds, and several state road rules also have restricted towing speeds. NSW and Western Australia, for example, are limited to 100 kph if the tow vehicle's GVM and/or GCM is over 4,500 kg.

Know the weight of your horses, drivers, passengers, and equipment so that you can work out the weights discussed above. There are also regulations and road rules regarding the length of the combined vehicle and float to have a "Do Not Overtake" sticker on the back left-hand side of the float.

All vehicles must have electric brakes fitted (or, in some cases, override breaks to a max of 2,000 kg) when towing a float over 750 kg. Check that electric brakes are adjusted correctly and can be adjusted for softer braking around town or heavier braking for faster highway driving via the controller in the tow vehicle.

Sunset at Zelper ASH Stud. Owner / photo by Sally Esdaile.

All trailer safety chains and 'D' shackles must be connected and rated to the correct braking capacity to take the sudden strain of the float if it becomes disconnected, especially with your horse(s) in it.

If your float body is wider than your tow vehicle, you may legally, and for safety reasons, need towing mirrors to help eliminate those blind spots.

A live animal in the float can move around so the weight can shift, and the horse can lose balance. Drive, corner, and brake smoothly and carefully.

Monitoring your horses while travelling is possible in the modern car and float setup. There are some excellent camera systems available at relatively low cost. Many are reviewed on the internet. In addition, your float manufacturer usually carries a list of the ones preferred by their customers and will set it up for you.

A UHF radio in your tow vehicle is also a useful safety device, especially if travelling on the highway. Tuned to channel 40, the 'highway safety channel,' you will hear of roadblocks, wide loads, and other useful information. A friend had her horse fall in the horse float, and she was contacted by a truck driver following, and she was able to pull over before any serious injury was done to the horse, thanks to UHF communication.

Printed with permission from the Truck Friendly Website:

https://www.truckfriendly.com.au/

Arakoola Lyric HSH and Arakoola Pride of Erin HSH, with the Hodges family. Photo Janet Hodges.

Chapter 32

Tools

HorseRecords

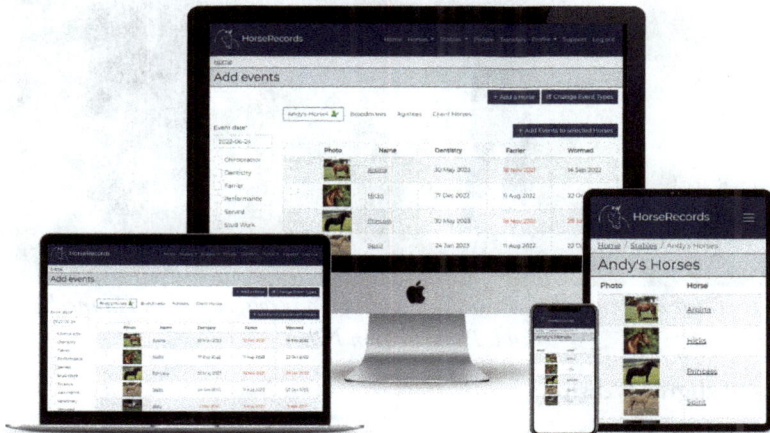

Once everything was a paper record, and our filing cabinets became bigger to cater for our needs. I still have every contract, with details of breeding and pedigrees back to the stud's beginnings in 1967. This has been an invaluable record when people have requested information. However, it has taken a lot of work to order, label, and file. Paper records deteriorate, and notebooks go missing. Many records are now spread on Email, Facebook, and Messenger, with no ability for such information to be transferred or recorded as historical data available to others.

My son-in-law, Andy, developed a wonderful programme which we use all the time. HorseRecords has been excellent for putting everything into one place and sharing it with other family members and clients. Then, at the click of a button, records can be transferred to an owner.

HorseRecords records all our pedigrees and progeny, photos, certificates and files, performance records, daily training, stud activity (such as scans), agistment records, and costs. It automatically calculates foal due dates and gives reminder prompts when events are due. It also has a free foaling date calculator and a names generator! It is flexible, so you can personalize it to suit your circumstances. When the vet is here, we can just look at any record on our mobile phones. Services are continually added according to feedback from clients. It is so simple. I don't know how we did without it.

HorseRecords: https://horserecords.info/

Masterpiece of Hannerley, Australian Riding Pony colt, owned by Hannerley Stud. Photo Sharon Nihot.

All-Breed Pedigree Website

I use this tool to access pedigrees and progeny that are not on a breed website. You may like to look up the names and pedigrees of horses mentioned in this book. Its accuracy is variable; however, I like to use it to record pedigrees of unregistered progeny from outside mares and for "hypo-matings". A hypo-mating is a projection of what a foal's pedigree would look like if you were to breed sire x to dam y. It highlights the inbreeding in the pedigree and for a small fee, can give you an inbreeding coefficient.

All-breed Pedigree Website: https://www.allbreedpedigree.com/

Your Breed Statistics

Your breed website is a most important tool to be used frequently. Familiarise yourself with it and point clients to it. Owners will have access to their association's contact details, important information about their breed's history, registration details, and many online tools, such as forms to print. Your breed's stud book, pedigree and often photos are available on its website. You can find the status of registered horses and whether a membership is current. It is the first port of call for newcomers to a breed. If you steer your clients to the website, you will add professionalism to your services and be seen as a reputable source of information.

Some breeds can give you excellent information about wins and earnings, especially in the racing industry. These are important statistics for your stud, recording background histories, and research. Top sire's lists, broodmare sires, number of progeny, performance statistics, and many other records are valuable tools. Make sure you have this information at your fingertips. For example, here is an excellent site for campdraft statistics and nicks:

Braeview Stud Campdraft blog: https://braeview.com.au/campdrafting-blog/

With breeds that don't publish statistics, you will rely more on breed magazines, stud advertising (which might be an embellishment or inaccurate), and books specifically on your breed's history. Collect books for your library that you can reference. If you record your horse's performances in HorseRecords, you will compile many of your own statistics, which will benefit you in the future, especially as you can print them out to give to clients.

Colour Calculator

It pays to know the colour genetics of your horses for colour predictions. Clients will often ask about possibilities. You can steer them to this useful tool from Animal Genetics:

https://www.animalgenetics.us/genetics/horseColorCalculator.html

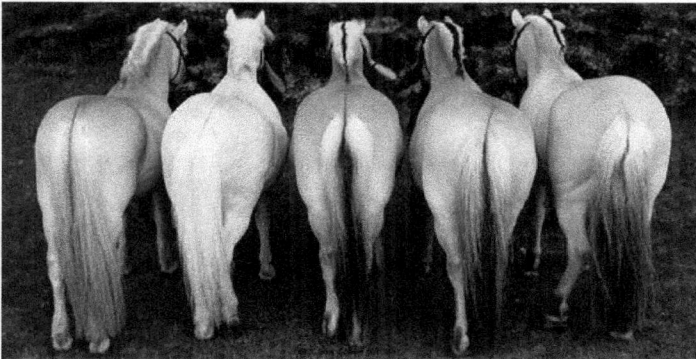

Norwegian Fjords, a type instantly recognisable. Public domain.

The benefits of the above tools will only become apparent the more you use them. Become versatile in their use to benefit your stud and your clients. These days YouTube is a way to gain basic education on any subject. It could be considered the university of the modern age. However, relying only on Mr. Google or social media, for information gathering, is unprofessional. Every tradesperson has his tools, and these are your tools. Convenience and professionalism allow you to spend more time outside with your horses and clients and less time in the office.

Bill Willoughby bronco branding from his QH horse Casper (by Great Stuff). This horse has won Year End Championships, Open drafts, Bronco Branding, only horse in Australia, that has won in every form of competition roping. Note the collar and rope. Photo Prue Chapman.

Bronco yarding around 1975. The horse pulls the calf to the ramp, at the panel, whereupon the rope drops into a slot. The calf is then pulled up to the panel and tied, to enable the stockmen to brand and castrate.

Chapter 33

Marketing and promotion

Can people find you? Do they know you exist? Do they know what you sell? How do you reach your market? Have you Googled your name / stud / breed to find who and what comes up? Are you *visible?*

Look at other stud's marketing. Does it look friendly and professional? What do you like and not like about it? Can it be easily viewed on a mobile phone? Talk to your mentors about what has worked or not worked for them. What are your competitors doing that you can learn from? What sets you apart? What are your strengths or weaknesses?

Good PR starts with the telephone. You need a phone to receive calls and an answering service with voicemail, *not* a "please leave a 10 second message." Let them know they have reached the stud (name) and that you will get back to them as soon as possible. When you pick up the phone, answer with your name or stud name. This is so important, yet often forgotten. Learn to answer the phone in a professional, friendly manner. Don't give the caller the impression they are time-wasters. First impressions count!

Every stud needs an email address for newsletters, contacting clients, etc. Have an email signature, including on your mobile phone. Keep a list of all your contacts, especially tradespeople and horse people. Over time, you can develop a big email list. Better still, have an email address you only use for your horse business. People prefer messaging and emails because they can do this at any time of the day or night when it is convenient.

A stud website is like a shop window. You can put as much as you like on there, even stud contracts, articles, or fact sheets. Even if you are a hobby breeder, it is best to have a website. Unless you have an eye for design, have it designed professionally, with plenty of photos of individual horses and their histories. A history of the stud with your mission statement and contact details are a must. If you can't update it regularly, make it generic, so it doesn't date. Look at other stud's websites for inspiration. Decide on a colour scheme and preferably add your personal branding. The finished product should look clean and inviting.

Of late, the popularity of a stud website has significantly decreased. Many haven't been updated in years, or the URLs are broken. I'm unsure if it's the rise of Facebook to promote and sell horses or if people are not perceiving the benefits of maintaining a website. If you take pride in its presentation, it asserts your horses as a quality product worthy of more than a quick Facebook post. It creates a more professional image.

A Facebook page is as important as a website. It too is a shop window but can be much more easily updated than a website. It is also interactive. Best of all, it is free. Horses can be listed for sale on a business page but not on groups unless shared from a business page. So keep your Facebook page separate from your personal page. Visitors sign on to learn about your horses, not your life's details and misfortunes. Remember, most will view it on their mobile phone so check for layout on each medium.

In every aspect, your Facebook page must look professional, with good photos, stories, or news. Keep it short, sharp, simple, and factual. Let the reader decide if something is "stunning" rather than you telling them so. Let *them* build an emotional connection. If you can post educational content, it helps build your readership and keeps existing followers engaged. Monitor it closely to keep out spammers and add positive client reviews. The number of likes or comments will tell you if you are doing it right.

If you allow messaging, answer messages promptly. Fewer people buy with a telephone call these days; mostly, you will receive a message. If you use Facebook infrequently, you are missing a major marketing opportunity.

Always return messages, return phone calls and emails the same day! I cannot tell you the number of times people have thanked me for a prompt reply. If other people don't do it, it makes it easier for us to shine. If you don't have the answer, let them know with a simple *"I don't know, but I will get back to you soon with the answer."* It's a common courtesy that apparently isn't that common. Horse people understand you may not be available immediately (you have an emergency, you're at an event), but you still need to get back to them with an auto-message, or at the worst, in person, the next day. An automatic turn off is to leave people hanging.

Performance

The number one promotional tool is performance!

Take your horses off-site at every available opportunity. People notice how you and your horses behave and how you respond to them. They notice and comment on horses they like. We have found that buyers notice brands and often find another "Chalani" owner simply by seeing a Chalani branded horse at an event. They can form great friendships and give excellent feedback.

Performance includes serious high-level competition, especially if your horses are successful in another ownership. Nothing beats success at National or Royal levels, annual Championships, listings on Annual Ratings, or high Lifetime Earnings. As they say, success breeds success!

Record and publish your horse's performances. This provides history for comparison and builds buyer interest. Buy *good* photos and post them about, but do not photoshop or alter the horse in any way. Clients can spot it a mile off and will wonder why. Potential clients watch out for the new up-and-coming stars and those who are successful. But it is also important for the older horses which may no longer be performing. Note: Always use your horse's *registered* name when posting, not his stable name, as people quickly forget!

Just as important is attending clinics, trail rides, competitions, parades, birthday parties, training days, mustering, or lending a horse to a Pony Clubber or a disabled rider's programme. These are all activities to post photos. Your horses (and progeny) must be *seen!*

> *"There are hundreds of stallions that get only a handful of mares each year, while there are a few that are popular. And so often, that popularity is based on three things: valid performance, correct conformation, and almost more importantly, not hiding his light under a basket."* – Denny Emerson

Networking

Step out there and talk to people. Network. Conversations with peers are important for insider information that is "hidden" from the average breeder, and know it gives a backup network for help in tough times.

One of the disadvantages of doing everything yourself and not taking your horses off the place is that you make yourself invisible. Your farrier, vet, and other professionals who come onto the place can be the best promoters of your name. Think of other ways you can have visitors come to you. Sell something on-farm, eggs, plants - anything which will have people coming in your driveway. Have your place tidy as if you are expecting visitors, including your horse's hooves, manes and tails. Have a sign on your gate.

Stud Open Days, clinics, gelding days, on-farm sales in conjunction with your annual field day or cattle sales and demonstrations, are all excellent to have your name promoted locally. Don't forget to tell your local paper. They'll likely send someone out to report on it.

Off-stud, it pays to become involved on breed committees, your local agencies, and support groups. Volunteer to set up or steward at local events or a disabled programme. Set up a blog on a topic that interests horse people. Write articles. Referrals and word of mouth are more effective than anything.

Set yourself up with a logo. It can be a photo or a design that reflects your stud. It may be helpful to have two that are a different shape. One for 'stamping' as a single colour, the other for glossy print media and larger printing options such as signage or banners. Stick to your branding colours and font styles. Make sure you have a decent business card (preferably with your logo) and hand it out at every opportunity. A most effective free advertising is the sign on your horse float. Banners are excellent for trade stalls, exhibitions, and shows. If you have a display, use a tablecloth in your colour, then add photo books that people can thumb through, handouts, and your business cards. You can even have a laptop with a video loop of stud happenings or your stallion.

Deciding on a slogan to reflect your goals is a good move. Add this to your advertising and on your banners. I saw this one in a vegetable shop and thought it excellent for our purposes: *Quality is our strength and your protection.* We have used it for over 40 years.

Think of ways you can have your name noticed with merchandise like baseball caps, jackets, saddle-cloths, glassware and mugs. In addition, your own wine label can be successful if given out as prizes, gifts, or sponsorship.

Media

These days, videos are used for education, information sharing, and sales. Set up a YouTube channel in your stud's name if you are doing videos. YouTube has plenty of information on how to set this up and tips for using it to your advantage. You can put your stud profile on Linkedin. These are tools that come up when someone Googles your name.

If you have writing skills, you can send information to newspapers and magazines about something related to your stud. Put "Press Release for Immediate Release" as the heading. Don't write it so that it sounds like an advert. Something like "Local stud achieves national success" is far better, as a headline with a photo of the winning horses, with a little story behind the success. It is best to make it sound like a story rather than just listing some facts. Finally, add your contact details so the reporter can contact you. It needs to be no more than a page/page and a half. If you cannot write something, ring the local reporter to see if the story is of interest. They may come out to do a photo shoot.

Everything comes with a trade-off. The less you can rely on paid advertising, the better initially. Paid advertising is hard to personalize and to reach your target audience it must be repeated for the best results, which becomes expensive. It pays to have it professionally designed to stand out from others. Think about your advert many weeks before you need to use it.

The key to good advertising is first impression impact. It doesn't need to be fancy. You don't need to copy other adverts because "that is how your breed advertises." Remember, you want to be noticed. It must not be cluttered but rather have plenty of space around photos and writing. Choose background colours with the photo in mind and the writing in keeping with your personal branding and colours. Choose one or two good photos, depending on the size of the advert.

Make sure you don't stand out for all the wrong reasons. Spelling errors and clichés are a major downfall; bad-quality photos or angles are very off-putting. Readers agree that the worst adverts have been photo-shopped or have the horse's photo "cut out" from its background. A typical mistake is to use a canter photo where the horse is disunited. Once you see it, you can't unsee it.

Advertising on the right-hand side page of a magazine is considered more visible than on the left-hand side. Sometimes you can combine with another breeder with horses of the same lines and book a whole page, where you might otherwise have had two half pages on separate pages. You can also book a section for local studs only or have a field day with other studs. Use your imagination. Is it creative? What sets your advert apart from the other 50 or so?

Regular classified advertising in your breed magazine is also effective and cheap. Support your magazine. Think of it as "sponsorship" by having an ongoing classified advert.

Another option is the Advertorial. You pay for an advertisement on condition they print a story on your stud or stallion. Usually, the printed article takes up as much space as the advert. You will nearly always have to submit the article yourself. Magazines will often ring you to see if you will advertise with them. This is your chance to ask for an Advertorial.

Repeat buyers are the icing on the cake. You know they have been satisfied and are often the best promoters of your horses. Give them discounts or something that shows your appreciation. Repeat customers are better than having to gain new ones all the time. Great customer service gives a good experience, whether clients buy or not. They tell friends or may come back later. Common courtesies are noticed. Honesty and good, timely service are the best advertisements one can get. Word of mouth is free and travels far and wide.

The best way to achieve what you desire is to hang out with the right people and copy them. Smart people don't try to reinvent the wheel.

Remember, whatever you do is not just a promotion of your name or stud but also your breed. You are a representative of your breed and your breed organization. Always consider how others see you. Your record, your integrity, and your professionalism are your reputation. It takes years to build and can be quickly lost.

"Success often leads to arrogance, and arrogance leads to failure. Ego is the enemy of successful marketing." - "22 Immutable Laws of Marketing" by Al Ries and Jack Trout.

Haydon Sun Charm HSH mare owned by Emu Gully stud. Photo courtesy Lynda Rogers.

Polocrosse ASHs - agility, courage, speed, stamina. Below: Circley Carnivale, HSH owned by Kylie Gould. Unlike polo, stallions are not permitted to play. Photos by Cathy Trail.

Chapter 34

Photographing horses

For the life of your stud, be aware of how important photos and media presence are.

EVERYTHING depends on your photos for advertisements, promotions, memories, and records. As someone who was married to a professional equine photographer, I can confidently say what is needed in the horse world is *good photos*.

Horses are particularly challenging subjects to photograph correctly. Your favourite photo may not be suitable, though it might be fine for showing your friends or passing on to an enquirer. Are you looking for a calendar/scenic shot, a newsy shot, a sales shot, or a promotional tool? Knowing what you want the photo for, you can determine what type will suit.

These days people often buy horses sight unseen, and they only have your photos and maybe a video to assist them. A bad photo or video can take thousands of dollars off the value of a horse or result in a no-sale. If you find the requirements for getting a good photo too difficult or your results are not up to scratch, then a professional photographer is the answer. Just make sure that the person you hire is experienced in photographing horses and your breed. Some photographers are excellent with action photos but don't get good standing shots. Others won't listen to what the client wants. Photographers experienced in taking photos at yearling sales and for magazines are generally the best in the business.

It is imperative that you never use an incorrect photo for advertising.

A bad photo will haunt the horse for the rest of its life. Once people see it, they won't unsee it. They just dismiss the horse from their radar. And don't try to retouch a bad photo. It doesn't work!

Just because you love the photo doesn't make it newsworthy; just as someone may love their horse doesn't necessarily make it breeding material. A photo that reminds you of a great time you had doesn't evoke the same response from someone who was not there. Such photos are great for the album but not for stud promotion. If you are trying to show news, you can get away with capturing the moment without it being perfect, so long as the horse is shown in a good light. These are the distinctions we must make if we are endeavoring to become more knowledgeable and professional in our approach to promotion.

Magazine submissions or uploading to Facebook, especially if a stallion is standing at public stud, represents an example of your breeding programme. This should be common sense, but some professional photographers and publicity officers will submit poor photos to magazines believing they are "nice" photos. I often take photos of my horses at an event, only to hear the click of a professional photographer taking it from the wrong angle.

The popular camera is the mobile phone. Many now have good zoom lenses and reasonable resolution. The downside is the delay in pressing the shutter button, which often means the animal has moved out of the frame entirely or flicked its ears back. The zoom is rarely enough to counteract the distortion resulting from being too close to the subject. The resolution is usually not good enough for magazine reproduction, though, with modern printing, even this is much better.

You can edit photos in your phone gallery, such as cropping, which removes the unnecessary background, or turn something into a head shot when the overall photo isn't good. You don't need anything more for Facebook, which gives you experience taking shots.

Cropping a bad photo into a good one with just one click on the "crop" feature of your phone or edit programme. Tooravale Indiana, Welsh B.

To obtain good quality photos of your horse, you need a good quality Digital-SLR camera with a zoom lens (minimum 200 mm). For the popular Western ¾ front pose, the photographer needs at least 200 mm of zoom! A fast shutter speed is also an advantage.

Use an SLR camera with the motor-drive setting and take multiple shots. That is one of the primary differences between an ordinary camera and a more expensive one. Test several cameras to find the one; ask the photographer at the events what they use and recommend. They love talking about their cameras.

Not all breeds are photographed in the same manner, and a photographer, like a judge, must have extensive knowledge of the breed so that the horse looks like a good example of its breed. Further knowledge is needed to work within breed guidelines and still accentuate an individual horse's good points and lessen things that are not. Photograph as close to the breed standard as possible.

The photographer must be many metres away from the subject so that he can zoom in to fill the frame. Amateur photographers note the zoom part. Anything less will distort the horse's shape, giving the appearance of short legs and a big head. Focus on the eye. If the eye is in focus the rest will be.

Play around with your camera and start practicing. Take heaps of pictures and select the good ones. It isn't that hard! Leave out the glitz and glamour and keep it simple. You can experiment some more when you have a good grasp of the basics.

The Stand-up Shot

If you want a good stand-up shot, you need a photo *session*, not something caught casually at a show. This is the classic advertising shot. It needs to be thought about before he goes into winter coat, and before the stud season if a stallion. You need a good background, a well-fitting halter or bridle, and a handler with a stick to nudge the horse into the right position. A third person to get the horse's attention is also necessary. Expect to take an hour; less is a bonus.

A good stand-up photo is more than just getting the horse's ears pricked. Think about your background, the lighting, and the position of the horse's legs. He needs to be on level ground and not look like he is standing downhill, with one end in a hole or grass that hides his legs. The handler should look neat and tidy or far enough back to be cut out of the picture.

Plan the shoot meticulously in advance. You don't want the eye of the viewer off the subject. The best photographer can only shoot what the horse will give, so train the horse to hold a pose long enough for photos. The handler's job is to keep the horse still, except when the photographer needs a foot moved or to change the angle.

Choose backgrounds which contrast and focus on the eye. Left Chalani Charm HSH. Right Chalani Jetstream HSH.

Presenting the horse well makes a huge difference to the result and how much money ends up in your pocket. Have the gear clean. Washing is essential, and so is trimming ears and bridle path. Fetlocks, mane, and tail should be trimmed, pulled, or prepared per breed requirements, and the halter or bridle correctly fitted. Is his tail the right length for the photo? Remember that his tail will look longer while standing, so you may need to shorten it a bit. You don't want his tail to hide his back legs. (Perhaps you do?).

Also, experiment with the bridle. The type of bridle, browband, plain/coloured, nose-band or not, can all make a difference. Have tidy hooves and oil them. The horse should look as if it's ready to parade before a judge because once you put your photo "out there," everyone becomes a judge.

The background must be clear, with no feeders, posts, poles, rubbish, sheds, vehicles, or animals lurking in your frame. A fence or plain wall behind can be useful for keeping him straight. When photographing coloured horses such as Paints or Appaloosas, the ideal background is a green grassy hill, distant mountains, or anything that provides a blank, unobtrusive 'canvas' for the horse's coat pattern. Even with solid coat patterns, you need something bland, not murky. It is essential to have a light background for a black horse. Use your imagination – a dam wall, the beach, a forest, a cliff face.

Early morning or late afternoon light is best as the light is on the side and not overhead. You can get away with other times if the sky is overcast. Look at your horse to see if there are any unwanted dark shadows along his side. Your back will be to the sun. If you experiment with backgrounds (use the dog), you will find a spot that works for that time of day and can be used and reused with many horses.

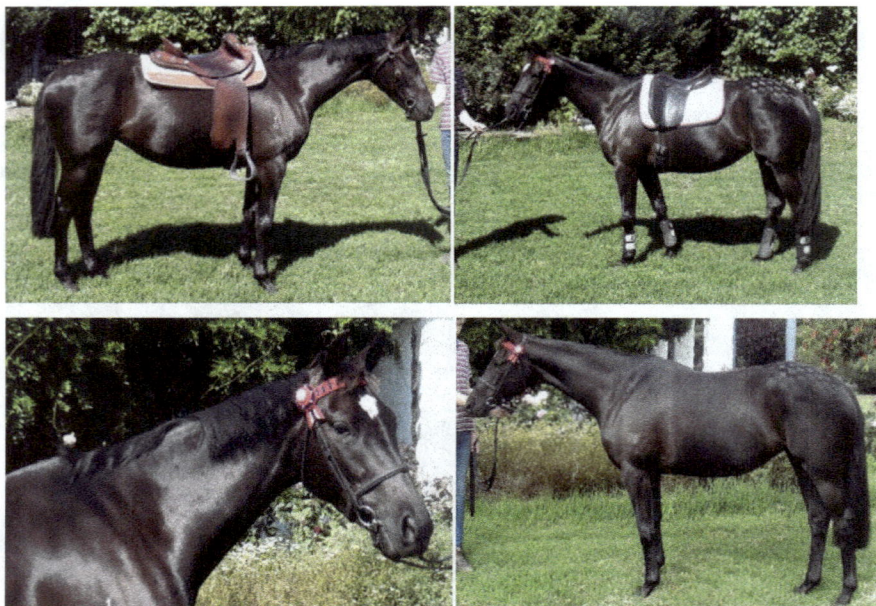

Ashborns Lady Brianna HSH. These photos show the same horse, with different styles, depending on your target audience. Once you have the right photo, you can crop out the handler.

You may be restricted by the light as to which side you take. You want the light on his face, so turning him the other way may not work. If you take head and shoulder profiles, make sure the light is shining on his eye so that it looks glossy. A head shot can look impressive against a dark background, such as a stable or shed. Some horses look better on one side or the other because of their markings or how the mane lies.

You can also experiment with taking photos with the saddle, with or without the rider. Some horses need the saddle to hide a long/weak back/poor wither. Others just look better, as it emphasizes their shoulder. If you want to make him look tall, put a short rider on his back or use a short handler, and vice versa. Check if he looks better with a shaped or square saddlecloth.

When is side-on not really side-on? The photographer needs to stand square on, level with the horse's *girth*. Too many photos have the photographer standing at the horse's forequarters and too close to the horse! This is the most common mistake of all. If the horse is weak in the hindquarters, it will look worse. When standing opposite the hindquarters, its neck will look too short, even if it is not! If the horse is not correct/square in the legs and hocks, do not do a ¾ front photo; only do a side-on shot. To enhance the hindquarters, particularly in western breeds, ¾ rear shots are done. If correct, they should show the side outline of the horse's gaskin. (See p48 or 75.)

Each leg must be visible. Two legs should never be together like a post. Ideally, the far side legs should be slightly inside the near side legs to make each leg visible. In some breeds, they will be more stretched. It is not as difficult as it looks if the horse is already well-balanced and will pose. If not, spend a little time repositioning his legs by moving him back or forth a step or a half, pressing your toe on his coronary band, or pushing on his chest with the knob of your whip. (This is where your prior stand-still training comes in, and your patience is tested.)

The photographer's role is to tell you when your horse's legs are in the correct position and advise you when not. If he doesn't, ask him. He is the one looking through the lens!

Have the "ear-pricker" stand slightly to the side where you want the horse to look (not directly in front of the horse). Usually, you'd like the head turned just enough to see the outline of the far side of the opposite eye. Be aware that his head can cast a shadow on his neck, which you want to avoid if he has a nice gullet.

The "ear-pricker" must have multiple tools at his disposal. We like to use a cap, rag, whip, squeaky dog toy (this is a good one to have in the pocket when you are on your own), pellets in a bucket, and an umbrella, to open and shut. Only use the last when you've tried the others and want to wake him up a bit. Keep fly spray handy.

The ear-pricker stands outside the picture and shakes or tosses various items to get the ears pricked. Sometimes picking some grass and showing it to the horse works. With a stallion, you can stand or walk a mare close to him. What works for one horse may not work for another, particularly in its home surroundings where all is boring.

Remember, the director of the show is the photographer. Prime the ear pricker not to do anything until the photographer says "now." Only when the horse is correctly positioned, ask for pricked ears. The horse must look alert, not sleepy. Focus on the eye before snapping the shot.

Take whatever photos you can initially, and work on getting better ones as you go. Periodically check what you have to see if the horse has blinked or swished its tail at the wrong time. Is the lighting and background working, or should you move him elsewhere? A little extra time is worth the effort.

Your effort reflects the value you place on your horse.

Action shots

You don't want pictures of horses in motion if they are on the wrong stride. It nearly always distorts the horse, makes the neck look shorter, the back hollow, or the hindquarters weak. If not taken correctly, the walk can look like the horse is dead or almost stopped. Click just as the nearest hind leg is furthest back because this is when the stride shows maximum reach, and the horse's front leg is coming forward.

Strides of the walk - Chalani Minerva HSH with Kim Ide.

The trot is the easiest to photograph. Just click as the rider is coming down to sit. Say "up/down" in your head and click on "down." That means the outside foreleg is on the ground, and the nearest one stretches forward. This presumes the rider is on the correct diagonal. If there is no rider, you must click as if, or slightly before, the nearest hind leg is the most far back to allow for your reaction time.

The back leg needs to be well behind when you take the trot and which one will vary depending on which side of the circle you are taking the shot from.

The canter must be taken when the nearest hind leg is under the belly or back, still on the ground, not when only one front leg is on the ground. You don't want to show a horse on the forehand. The best way to do this is to count one, two, three – one, two, three, and click on "one." Other shots are all four feet off the ground, or the stretch, where the foreleg is at a maximum stretch before being placed on the ground. Never put up a photo of a disunited horse or one on the wrong leg!

Set your camera in advance on "multiple burst" shots. It is called motor drive. Hold the shutter down to take three to five quick shots in a row. You will usually find one of them has snapped the correct timing. I have even held it down and snapped a whole workout. With digital, it is simply a continuous shot/drive instead of a single shot and is measured in frames per second. Digital is hard to shoot action any other way because of the delayed shutter action before you can snap the next shot.

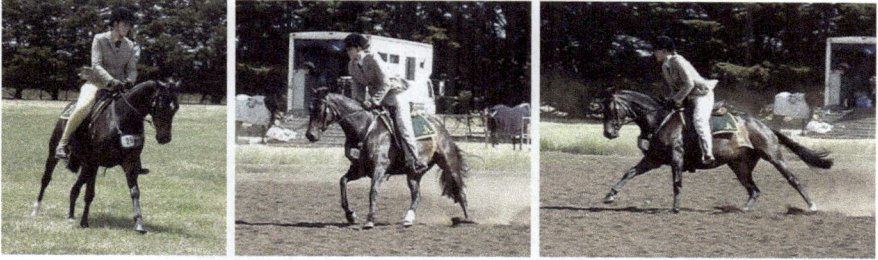

Once again, if you click when the nearest hind leg is back you will take the best shots.

Mood photos

These are taken for impact (calendar shots or advertising etc) with the aim to create an emotive response by the viewer. They achieve their aim simply because they are more of a rarity in the advertising world, but they need a little imagination and creativity. They often involve breaking the rules for lighting or background. But that doesn't mean you can't take them; you just need to carry your camera / phone around with you as often as possible so that you can capture that indescribable "moment." Remember: take it from a distance and zoom in! You never know, you might just capture a masterpiece; at least you will have a great memory.

Joseph's Dream Appaloosas, Namibia, backlit. Photos Joseph's Dream.

This photo of Rannock was used in all our advertising. It always attracted attention. It is the classic "look of eagles" head shot designed for impact. It was taken a long distance away while he was out running with his mares to capture alertness, and zoomed in to avoid "boof head" syndrome, which blurred the background against the sky. You should only do headshots if the horse has a good head in the first place. Photo Peter Gower.

Presentation shots

The temptation when doing presentation shots is to get up close to the sash or ribbon. Remember to position yourself back at full zoom so you don't get "boof-head" syndrome. Is your back to the sun, and do you have the clearest possible background? Place the sash around the neck near the shoulder and pin it at the chest, so it sits flat. Position the ribbon the right way up, showing the event and date. Have an "ear-pricker" and the fly spray handy!

In this shot, I placed my right hand on the the horse's withers. This allowed me to stand far enough back to allow the horse to bend his head around *me,* without hiding the garland, so that the photographer could concentrate on taking the head from a suitable angle.

The author with Chalani Sunstream. Photo by Kangra.

Bad photography

Few photographers seem interested in taking a proper side-on shot of a horse at an event. This is the photo "standard." All else is a bonus. The three or two-legged horse with hidden legs creates unflattering optical illusions. Anywhere a horse is photographed, particularly if it is a stallion, is a "stud" shot. Handlers must pay attention in case the photographer is shooting his horse. Maybe its ears are back, or it is resting a leg, or its leg placement is wrong. The photographer should point this out.

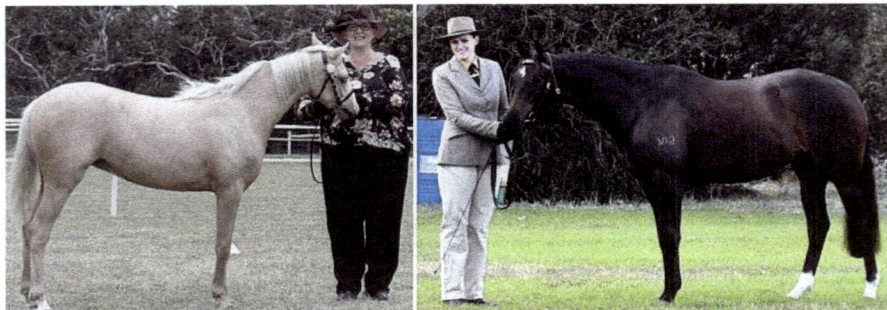

What is wrong with each of these photos?

Horse owners, particularly breeders, don't like to wade through hundreds of proofs and find hardly anything suitable. Photographers, please take fewer photos and set horses up better. If I want to take 1000s of photos and find something usable in the batch, I can do that myself, very economically. Don't create a two- or three-legged horse! No ridden horses on one foreleg and three in the air. No bad photos on your websites! If customers give feedback and it's not what you want to hear, please don't quote the "hundreds of people" who have bought your photos.

Unfortunately, a photographer is sometimes an "exclusive," and competitors don't get a choice. Owners may be forced to buy a second-rate shot as no other photographer is there. They consider themselves "lucky" that a photo was taken at all. We do have a choice, and that is not to buy.

Video

Sometimes it is easier to take videos than still photos. The same rules apply to lighting, background, and time of day. You need a handler to trot the horse out. The simple shots are walk and trot out and back, both in a straight line, to and from the camera. Then put the horse in a lunge area or arena. Using the lunging whip, set it loose to trot and canter around or even jump. You may need to have someone wave an umbrella to get sufficient activity. This is an excellent way to take a video of a foal with its dam.

If a youngster is trained to do obstacles, hose down, or self-load onto a float, take a video of it. If he is ridden, walk, trot, and canter as a minimum. Even better if you have a series of videos from different outings over time, especially of workouts. Vary it. You can even "snap" still shots from your video once you have downloaded it, picking the best frame. Film plenty of action on video and casual footage of happenings around the stud, as this can be used in promotion later if desired.

Join different clips together and put them on your YouTube channel. Any video for sales should be short, no more than five minutes or so, preferably joining content in chronological order. Printed information and contact details should be at the beginning so that a person doesn't have to go through the video again to find it. Then add the link to Facebook or interested parties. Don't try to learn the technical side immediately, and put yourself off forever. Instead, find someone who can assist you. You can edit your videos for free on the Microsoft platform Clipchamp: https://clipchamp.com/en/video-editor/

The correct way to stand a horse up when you can't photograph him from side on. The photographer has moved to the front to avoid a "three-legged" horse. All four legs are showing. Chalani Sunstream and Kim Ide. Note: Kim is standing in "formal" position.

Answer to Question:

The first photo shows the correct background and lighting on an overcast day. The photographer's position opposite the girth is correct, but the horse's legs are not properly positioned. As a result, he looks like a two-legged horse. The second horse has its legs correctly positioned for a side-on shot, but the photographer is too far to the front, opposite the horse's neck; hence he has snapped a three-legged horse. The background is not suitable for a dark horse, though that sometimes cannot be helped at an event.

I would be really interested to know how you go taking your own photos after reading this. I am happy to critique. Just email me with PHOTOS in the subject line. Do let your friends know.

Dollar, Matt Donoghue and Tomali Master Herbert ASH, Kate Growden, Adelaide Royal 2017 steward's horses. Photo Horizons Photography.

"Most of our job is carried out on bitumen or a sloped hard sand track, in all weather and includes periods of up to an hour standing in the same spot while classes are judged and presented, surrounded by grandstands, amplification and large crowds (all easy for an experienced horse). No day is ever the same with the atmosphere, crowd or sights. Heavy harness, light harness, showjumping, grand parades, loose cattle, school goats, all types of horses, section fours, working stock horses, motor bikes, Ute's, ignorant public, prams, umbrellas, ponchos, leading imported warmblood showjumpers, standing beside a nervous hack, leading ponies over drains, lights and sound effects, are all part of the daily job, not to mention nine nights of large fireworks displays etc. It is a most rewarding job but very hard on the horses."

– Matt Donoghue

Chapter 35

Buying and selling

In the horse business, you are in the business of SELLING. If you dislike parting with your horses or are indecisive about who should buy, don't breed. Never breed more than you are prepared to sell. Are you prepared to be a salesperson? Are you a people person? Effective sales need people skills. How many sales per year do you need? At what age will the horses be, and what level of education?

When should you sell a horse?

- Someone else can do better with the horse.

- It is no longer a good fit for your programme.

- To showcase your bloodline through performance by someone else.

- The horse is worth too much to afford to insure it at its present value.

- Retirement (horse or you) or change of direction (goals).

- It is no longer a breeding proposition.

- You sell everything you breed.

- You see the market falling away.

Never sell a horse simply because it is aged, useless, or frail. Take full responsibility for giving it a peaceful end so that it is not passed on or ends up in a dogger's pen.

"Once that horse leaves your place, no matter what the buyer assured you, there are zero guarantees. And I am not talking about bad luck, illness, injury, because those issues can happen to anyone. I am talking about the horse falling into poor management, tough training techniques, things like that. I don't mean that there shouldn't be horse sales. I am not a believer in having to keep a horse forever once you own it, if the horse is less than a good match. But we know the saying "caveat emptor," Latin for buyer beware.

There should also be a saying, "seller beware." Not everyone will take as good care of a horse as we might hope." – Denny Emerson

If you live a long way from the market, have a higher-value horse, or feel you are not much of a salesperson, taking it to a breed sale may be a better strategy. Many high-end breed sales offer incentives, competitions, and exclusives just for being entered into the sale.

Choose a sale that best fits the type of horse you are selling. Read the entry conditions very thoroughly before you commit. When are the closing dates? What will your reserve be? Will the cataloging fees, transportation, and other costs be prohibitive if you don't make your reserve? What disclosures, information, and extras, such as photos, do you need to provide? Are registrations current? What arrangements will you make if the horse doesn't sell?

Most breed sales require a pre-sale inspection by the organizers for identification, presentation, and suitability. Some will require current x-rays and health certificates. Who will display the horse for you, and how will it be presented? Have a blurb ready. Hint – use large font. Don't leave it to the auctioneer to invent something. These are all added expenses and logistics that you will need to factor in.

Buyers want to view the horse, sometimes several times, before auction day. Many will contact you before the sale seeking extra information. You must be easily contactable. Ensure you have your information handy and promote your horse well before and at the sale.

As a stud, you are not just selling a horse. You are showcasing your stud's name, brand, and reputation.

Deciding on a reserve can be quite complex. How do you compete in a market against established studs that produce top horses and are well-known and respected? What are others of similar breeding, age, sex, and style fetching? Are they top of the range, with a name trainer from a name stud? If this is the case, it is easy to set a lower price. If you are at a distant location, you may need to offer incentives, like transport discounts or other benefits. Don't put an exorbitant price on your horses only to have them sit in the paddock for the next 12 months. Holding out for the best price may be a mistake when the horse could drop dead tomorrow.

Have you ever analysed a good advert, seen something that grabbed you, and thought the advert ticked all the boxes for good advertising? Different styles may be required for a noticeboard, Facebook, website, classifieds, a local newsletter, or a major magazine. Knowing your market and placing the advert in the right medium is critical. A well-thought-out advert is part of your overall promotion. You want a nice clean advert that comes across as genuine. If the advert is for a magazine, the magazine can design one for you. They can even print off flyers from the advert for you.

Think of a suitable heading. Put the horse's registered name, breed, and how it is bred on the advert – or a link to its pedigree. A link gives the pedigree without clogging up the advert. If the price is non-negotiable, say so. If the horse is heavily discounted, say why. Add the location as many buyers will not buy from interstate. When you do any advertising, make sure you are easily contactable. List your number and/or email. If you're only available between certain hours, say so. Buyers may easily give up if they don't get a timely response.

RIVINGTON THE CATS PYJAMAS

Superb Childs Welsh A Gelding
11.3 3/8hh Bay 3yo. Reg WPCS, APSB & SHC.

This wonderful pony has the most fabulous nature and quiet temperament. Broken in by 11yo boy and has been perfect for kids right from the start. Ridden only by children, he is very straightforward and easy.

Soft snaffle mouth, goes kindly on the bit or on a loose rein. Steady, soft paces with a super smooth canter and great brakes. PJ will be an excellent First Ridden and Leading Rein Show Hunter pony, as well as an ideal pony club, dressage and interschool mount.

PJ is easy to catch, float and clip. He has done obstacle courses, trail rides with little jumps, lake rides, youngstock shows and has started his newcomer ridden year. He has been taught to have a gate opened and shut off him. Saddle up and enjoy a pony that is the same out and about as at home and doesn't require riding every day to maintain him. PJ doesn't care about being left tied to the float or where his mates are when being ridden.

PJ has been afforded the same training as our stock horses and he is just the most delightful loveable character that we have owned since a weanling. Unfortunately my boys are too tall and they have made the decision he's ready for a new little rider.

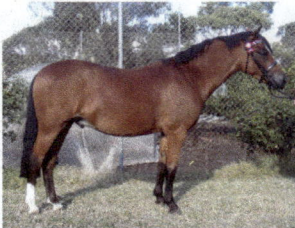

A good type of sales advert with the buyer in mind.

Make it easy to navigate, as you may only get a few seconds of interest before the person moves on to other adverts. The beauty of a well-laid-out website advert is that the person can do as much or as little browsing as they wish. You want to capture their attention and let them "discover" as that builds excitement. You can share a website advert on Facebook or do a Facebook advert with a link to further details on the website.

It is best to use one good photo with a link to further photos or videos, if you have them. A common mistake is using too many photos because you can't decide which is best. Be selective.

Videos make it easier to see what the horse is like, especially in movement. Have it work under saddle if broken, and show it led, towards, away, and side-on. Make sure there is no disunited canter, missed flying changes, mouth opening, head shaking and tail swishing. If your horse can't do more advanced things nicely, then just do one simple, clean video. The more expensive the horse, the more time you should put into doing a nice video. I suggest only putting in information which will not date, in case you don't sell the horse and want to use it later. *Leave out the bells, whistles, and fancy music and just put the facts.* Any further information can be on the post.

If it is not possible to do it yourself, you can go to the Fiverr website, and have a professionally made video for YouTube from your own footage which includes voice over, for a once off fee of around $100.   https://www.fiverr.com/

To stay in business, you must keep up with the competition.

Rules for selling:

- Present well - both horse and advert.

- Only use a good photo or video.

- Emphasize good points.

- State the facts.

- Add your contact details.

- List the location.

- Set the right price.

- Suggest what type of buyer is the best fit.

- Remember the horse must be sold as "fit for a purpose."

- Be clear and concise, without gimmicks or cliché.

- Use a simple layout/design - less is more.

- Don't sound desperate.

- Put yourself in the buyer's shoes.

A successful advert will result in a sale, but a non-sale doesn't mean the advert didn't reach an audience, just that no one was interested in buying at that time or at that price. So repeat the advert several times before changing it or listing it in different places.

A bad advert has turn-offs that the advertiser is unaware of. It may be poorly thought out, cluttered, or not well positioned. Poor spelling, grammar, and photos are the number one turn-offs; you want to be seen for the right reasons. Don't place an advert when you are unavailable to answer queries.

Another turn-off is the price on application (POA). Many people assume that the price will be out of reach. Better to list the price or provide a link to further information which includes the price. POA strikes others as shifty, as though the horse would be one price

Joseph's Dream Appaloosas, Namibia. Photo Joseph's Dream. Quality resonates with buyers.

for one buyer and a different price for another. Or that the seller would sell me the horse, only to renege if someone else offers more. If you are unwilling to state the price, put "best offer over x" so buyers have a ballpark figure. Or "price negotiable."

Never write things like "no time-wasters," "no photo-collectors," and "serious inquiries only." Essentially, everybody is a time-waster except for the person who buys. A used car salesperson expects someone to shop around or not get back to him unless they are interested. Why should selling a horse be any different?

A person may inquire about different horses with clients in mind, just scouting for others. Then, if suitable, they will recommend the horse and pass on the details to the owner to take it further. There is nothing wrong with a seller giving the enquirer a 24–48-hour deadline to get back to them, to allow thinking time, and offer further information.

Expression of interest (EOI) is another term. It doesn't guarantee that the buyer will get it if others also express interest. Deposit money is not usually required for an EOI, but if it is, it is fully refundable, as there is no guarantee who will be successful. The seller wants to go through a process to check out potential buyers. It weeds out time-wasters but allows everyone to apply by a certain date. It removes the pressure of the impulsive or harassing buyer who may not be the most suitable.

EOI is sometimes used for foals, or foals in utero, where the intention is to find serious buyers who will collect at weaning. We often do a payment plan under these circumstances, with full payment completed at weaning. In the contract, list the handling, registrations, gelding, and vaccinations you will provide the buyer. It works both ways, as the prospective purchaser may also be looking for a specific gender, height, or colour. It means, "This foal may be for sale. Contact us now to be given the first opportunity to purchase."

EOI is also used to sell a child's pony, with several people on a waiting list. The seller then has a documented list of people with names and contact details when the pony becomes available.

To ascertain if a potential buyer will provide a suitable home, listen to their questions and put yourself in the buyer's shoes. *Most people are not experienced buyers.* They may ask what might seem like silly questions, but you will soon get an idea of their experience by the language they use. If they haven't already told you, ask:

- What is your background in horses? (One horse owner, getting back into horses, trainer, etc.)

- What have you done with your horses? (Played polocrosse, dressage, trail riding.)

- What do you want to do with the horse? (Compete at riding club, breed a foal, sporting, show in futurities.)

- What supports do you have? (Work full-time with horses, lessons locally, mum rides as well.)

- What facilities do you have? (Agistment, horse float, fencing, acreage.)

- Who will be the rider? (Professional, keen child, nervous adult.)

- Can they supply references? Or a contact number you can ring to check suitability?

The buyer will likely ask about the history of the horse, its suitability, any illnesses or lameness it might have had, if a vet check is welcome, when it can be viewed, and if the price is negotiable. They may ask why you are selling. Anything really that hasn't been put in the advert. They will tell you about the rider experience if it is a riding horse. They may ask about transport possibilities and if any equipment comes with the horse. Someone who asks many questions may be doing so to establish trust and integrity, not just exploring the horse's suitability. The purchaser may seek the names of others who know the horse and can vouch for it. If you realize the horse is unsuitable, politely say that and thank them for their interest. Let them know of someone else who might have something suitable.

A stud visit is usually done to inspect the horse's suitability, condition and relations. A buyer may bring a friend or instructor. Have them leave their dogs in the car! If you have a ridden horse, ensure a rider is available with an enclosed space. Let them see the horse in its natural environment, catching, leading, and saddling up, then let your rider show it working. Putting the potential purchaser on the horse first up is not wise. Know that your public liability insurance is up to date, and only let the viewer ride with helmet and boots!

Vet checks are usually done at the buyer's cost after receiving a deposit. The deposit is refunded if it doesn't pass. Some may require x-rays. To avoid conflict of interest, give the purchaser the names of several vets in the area so they can choose one and explain their requirements. The vet will give the seller a form to sign confirming the horse has not been drugged, etc., before the examination. Note: drugging can only be ascertained by swabbing or blood testing, so this needs to be done immediately upon purchase if anything is suspected.

Once a sale has been agreed upon, you should require a deposit and provide a sale contract, unless the full amount is paid immediately by electronic transfer. Let the buyer know that the horse is not theirs, they cannot take the horse off the place, no transfer will take effect, and you can sell the horse to someone else, unless your conditions are met per the contract.

Leasing is another option suitable for studs, as is a "foal for a foal" arrangement. The contract should explain both party's responsibilities and what should happen should the mare die or other contingencies, such as need for surgery, and insurance. Handshake agreements are fine until something goes wrong that you didn't anticipate.

Rules for Buying

Inverse everything listed in Selling to understand the buying process.

- Be polite.

- Ask lots of questions after reading the advert in full. (Best if you have a checklist of your needs, including height and location.)

- Ask if the registration is current and all fees are paid.

- State the reason/purpose you wish to buy.

- Do your research – don't believe everything the seller says. (They may not know.)

- Search Google/ Facebook with the name of the horse.

- Ask what genetic diseases the horse has been tested for and what the results were.

- Don't exaggerate your abilities or experience.

- If an advert has a phone number, you are expected to ring. If an email, then email the seller.

- Arrive on time. If a cancellation is necessary, rebook with an apology.

- Don't expect the seller to keep a horse for you, so ask what the conditions of sale are and whether you are first in line.

- If you're buying sight unseen, try to get videos, referrals, and information from past owners.

- If it is a breeding horse, get a breeding soundness examination and a clear swab. If it is a riding horse, get a vet check.

- Don't drag the process out unnecessarily.

- Remember you don't own the horse until you have paid for it (with cooling off period).

- You can't return a horse just because you have changed your mind. (Consumer Protection Act.)

- Make sure you understand if a deposit is non-refundable or not. If you pay a deposit in cash, ask for a receipt.

- Ask for a contract of sale, especially if you are paying on a payment plan.

- If the horse is a giveaway, buy for $1 and obtain a receipt as the new legal owner. (Otherwise, you risk the owner demanding it back later, claiming it was a free lease).

It is not hard for you as a buyer to say, "I'll have a think and let you know in 24 hours if I want to go further." Don't leave the seller hanging. And thank them for their time in showing you around. In a hot market, you may need to make a snap decision, so this is why your prior research and market knowledge means everything.

Contracts

Contracts are important for almost any interaction with clients these days. It is the professional standard. You have too much at stake without one for leasing, selling, payment plans, services to your stallion, foal arrangements, etc. You can change a contract to suit the individual client. It doesn't need to be in "legalese." If you are buying as a partnership or syndicate, have a written agreement about what should happen if the horse is sold and if a syndicate member wants to sell his stake. Who will have the first option? What happens if the horse dies? Must it be insured? Have a contract if it is a giveaway with conditions. Better to sell for $1 if it is without conditions and give a receipt for the sale.

Never offer a payment plan on your cheaper stock. This is a red flag. The purchaser either has the money now, or there is no sale. They can pay it off to their lender, not you.

Make sure anything you said on the phone is in the contract if you both agree to it. Otherwise, have a clause which says, "This contract supersedes any prior understandings." That way, you are protected from someone coming back at you who says, "You said........". You can put your standard contracts on your website, so mare owners and buyers can download them, saving you time and unnecessary communications.

To develop a contract for a particular purpose, check what others have written or discuss with a legal person.

Difficult Clients

If an issue with a client arises, don't let your first response be defensive, accusatory, or denial. This only provokes the client to become more aggressive. Instead, take a deep breath and say something like, "I am listening to you. How do you think we can fix this?"

This approach usually defuses the situation enough to get some reasonable dialogue. Clients are often emotional and don't know what they want. Ask what *they* propose to resolve it to their satisfaction rather than you first offering a solution. Keep it confidential and request they do the same.

Then say, "Put it in an email, and we'll take a look at it/we'll see what we can do." This is not a promise or a commitment but places the onus back on them. It puts everything in writing! It gives you thinking time. Record the date and details of the interaction.

Often the email never comes, or it can be resolved simply by referring to your contract. Always reply in writing. You might make a counter-offer or indicate if you are willing to negotiate. Have someone else read it before sending it. You want it to be matter-of-fact and accurate, not emotional. If something is unlikely to be resolved simply or causing you a lot of anxiety, go straight to a lawyer for advice. They will usually give the first session free. This is why everything must be in writing.

Elephant circus costume. Springtime Penny Royal, Australian Riding Pony with Kim Ide.
Photos prove your horses. Let people know!

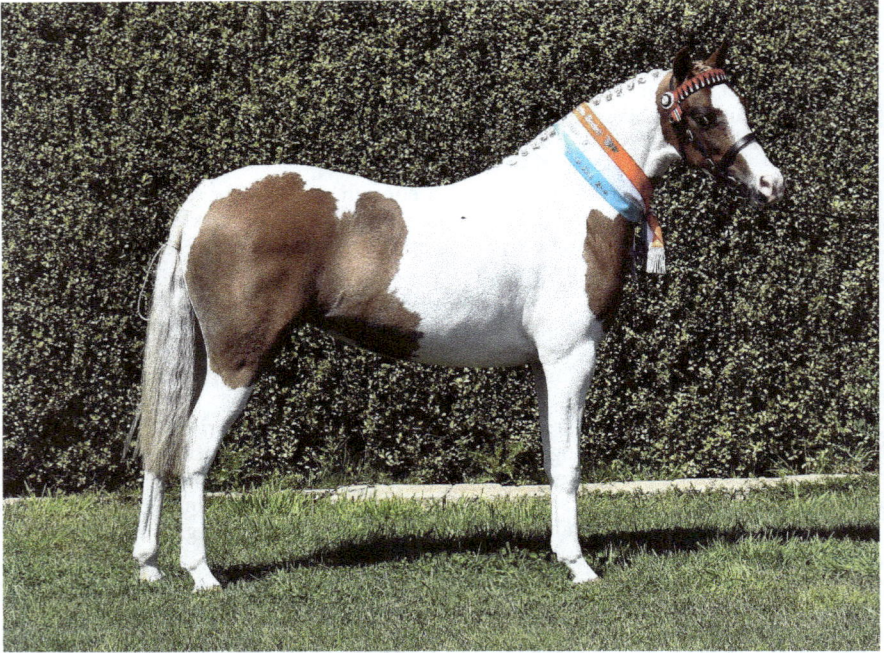

Exceptional presentation. Kyabra Park Ruby Rose, Arabian Riding Pony - Photo by Tegan McKenzie.

Three full siblings, L-R: Aloha Stars n Stripes, Aloha Aquarius, Aloha Silhouette. Photo courtesy Aloha Arabians.

Chapter 36

Preparing a horse for show or sale

By now, you have trained your horses well on the ground, they have manners, and you'd like to participate in a young horse show. The purpose of taking your horses to breed shows is to give them experience and for you to showcase your stock. Your training prepares your horse to stand for the judge or a photo and to confidently sell, knowing its qualities. A well-prepared horse will often sell simply because it has been seen!

Some would argue that breed shows are a disservice to the industry; it's about who is leading the horse, points, and ribbons, none of which counts because it is all subjective. It doesn't matter how nice a horse is or how many ribbons it receives. It doesn't guarantee that someone will buy it and train it to Grand Prix or the Tom Quilty endurance ride. Even if it does happen, it can take a decade or more. But a horse show may be the only way to draw attention if you sell youngsters.

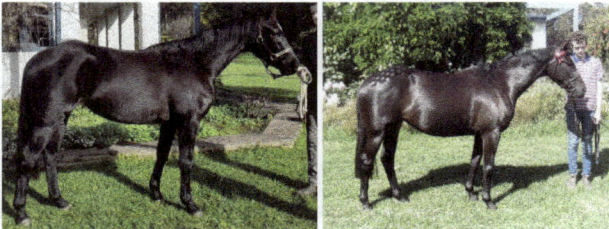

The difference a little maturity and conditioning makes. Ashborns Lady Brianna at 2 years and at 3 years.

If you are going for the experience, it doesn't mean you shouldn't be competitive. It takes a minimum of six weeks to prepare the horse. Stand back and look at your horse. Have someone lead him and trot him out, so you can appraise his movement the way the judge will. Does he need corrective trimming? What condition is he in? To have your horse in show condition, he must shine from within. To reach this goal, your horse must be 100% healthy.

The first thing is to work on a worming and feeding programme. Check the feed label for ingredients that your horse requires. Feed hay generously to bulk him out. That is important for traveling, as he may tuck up. Specially prepared oil products, daily canola oil, or beetroot powder can produce a shiny coat.

Before you rug your horse, wash him. Use a hood or a combo rug. Horses are measured from the centre of the chest to the rump. A *correctly fitting* rug of the right size should not rub your horse anywhere. Even so, some rugs will still rub. Rugs with a pleat on the shoulder reduce rubbing, as will a satin-lined bib. If a front strap rubs, thread it through an old woolen sock.

If your horse is in the sun during the day, his colour will fade, and his coat will dry out where he is not covered so use a cotton "sheet". Wash the sheets when they get dirty or sweaty. You may need several sheets to swap them for washing. In uncertain weather we use the shade-cloth style rugs.

Consider putting the horse under lights, or clipping it, if it hasn't dropped its coat. *Always* wash thoroughly before clipping! If you clip after the new coat begins to comes through, you can clip off old hair and not get the "clipped look." Greys, in particular, can be done up to the day before the show. Speak to your professional clipper to arrange the best time to clip out your horse. Add extra rugs, and beware the first time you put the saddle on, as it may feel quite different to the horse!

After four weeks, your horse will look and feel pretty good. If he is becoming a bit of a handful, then it's time to reduce his rations or increase his exercise. Regular exercise is necessary to get a nice muscle tone. If he is stabled, give him plenty of turnout time and protective boots if necessary. Young horses are prone to splints, so get into the habit of putting boots on your horses before you lunge or ride them.

Shoe and/or provide corrective hoof care as necessary. He doesn't need to be shod, but his feet must be in good order. A judge will notice this *first*, and about 50% of all presented horses are let down due to poorly trimmed hooves or crooked legs. Therefore, take a critical look at his legs and how he places his feet, as he may need corrective trimming to place and track correctly. Have the horse's feet trimmed/shod at least a week before your event, thus avoiding any chance of tenderness.

Tail care makes all the difference in your overall presentation. This commences months prior to your event by making sure it is not chewed off, by a calf or foal! Step back and look at the shape of your horse. If he is a long horse, cut his tail about chestnut length to give a shorter appearance. A shorter horse, therefore, requires a longer tail. Trim his tail midway between hock and fetlocks. If he carries his tail too high, you can add weights on the day.

The easiest way to keep your horse's tail clean and undamaged is to have a tail bag velcroed onto his rug. Another way is to place a stocking or tail cover over a clean, plaited tail or a tail bag plaited into his tail. .

Always wash the tail with shampoo and conditioner at the start of your preparation, and at least weekly thereafter. Do this again before the show. If the tail is white, it will need extra treatment with shampoo designed for white tails. Sprays such as "No Knots" are ideal for brushing through the tail with an appropriate brush that doesn't break the hairs. Always brush the bottom of the tail first, gradually brushing through to the top. Tails that are sun-bleached are best given a hair dye treatment about a week before the show with a human hair dye of the same colour as the tail. Wrap it in a plastic bag so the dye doesn't splash over the horse, and leave it for 20 minutes or so. If his coat looks particularly dry, you can give him several hot oil washes leading up to the final day. Coat colour "enhancers" are also available to lift the coat's colour.

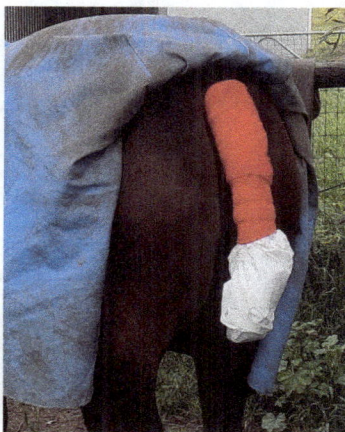

Hair dye wrap on tail

Make sure you have a helper lined up well before the show. He will be your groom, runner, or go-to person.

If you are doing led (in-hand or halter) classes, practice at home. Your horse should be obedient, quiet, walk out, trot out, and stop as asked. Teach him when you are in a "formal" position, to "pose" or stand without fidgeting. Have people walk around him like a judge would do. This takes time, but it is time well spent.

What is formal position? This is at attention in front of your horse, slightly to one side, and watching down the line to the judge. Here you have a clear view of your horse and the judge, merely with a turn of the head. As the judge walks past, swap sides so that you are still facing him. The horse is not to move when you are in a formal position. "Informal" position is when he is not posing, but must still stand *still*, such as when you are talking to someone, or you are grooming him. He can turn his head to look at things, or eat, or rest a leg, but he is to remain stationary. In this position you are relaxed and usually by his side.

Decide if you will show in-hand in a well-fitting bridle or halter. If a youngster hasn't been bitted for long, you can put a noseband on him, loose enough to put your fingers under the nosepiece. Then if necessary, you can stop and steer from the noseband rather than use the mouth. A lead rein with a divider is suitable for youngsters who are not yet confident in a bit. It is also helpful for those horses inclined to nip or grab the reins when bored or anxious. Practice with it. If you are using a western halter, you might need to add a light chain across the nose for extra control. I don't like the show "slips" which tighten up under the chin when the horse pulls against it, because a horse is too inclined to jerk its head up when any pressure is applied. This is not a good "look" in front of the judge.

If taking a stallion/gelding, make sure his penis is clean. Practice backing him up. On the day of the show, a sharp backup is usually all that is required to put it away.

No longer than a week before the show, clip out any white socks, as white hair always looks thicker than its counterpart and often won't clean up as well. Remove feathers unless your horse is a "feathered" breed. Tidy the mane for plaiting or remove it altogether. Introduce

Chalani Card Tricks HSH with Kim Ide. - Photo Lisa Gordon Photographics.

the sound and feel of clippers early in your show preparation. Don't leave it to the day before the show to introduce your horse to electric clippers!

The day before the show

Clip any final areas, including fetlocks and under the chin and ears. You can do this with small, sharp scissors if you don't have clippers. You don't need to clip into the ears but remove the fluff by closing the ear and snipping downwards. A grooming razor helps trim the hair around the nose and muzzle. (Check if your show rules allow this). Trim the tail carefully while holding it at its natural height. A little bit at a time is better than too much too soon. Next, thoroughly shampoo the horse's body and tail. There are shampoos specific to your horse's colour. These work to enhance natural colour. For white, grey or palomino tails, use purple shampoo for white tails

After rinsing out the shampoo and conditioner, scrape the horse down and spray finishing polish all over the body and through the tail, except avoid areas to be plaited and where the saddle lies, as it becomes slippery. Rug your horse up with clean rugs, hood, and tail bag. If you have lightweight leggings, put them on too. This will save time the next day if you have clean legs to start with.

If you are plaiting, plait up the night before and put a lightweight satin hood over the top. Spray the plaits with hairspray to hold them. You can buy false plaits if he has rubbed out a section.

Have your gear polished and clothes ready for the morning. We have a generic gear "checklist" for our shows/competitions. It is invaluable. It is easy to cross off those things not needed on the day and tick the others when packed. Check your grooming box to make sure you have everything you need. It is easiest to keep a separate grooming kit in your float just for shows. Pack your programme, registration certificates and double-check what time your first class is to allow plenty of time, especially if you are traveling a distance.

If you would like a copy of our checklist to alter for your own purposes, email me with CHECKLIST in the heading, and I can send it to you.

The day of the show

Upon arrival, take your horse off the float and give him a hay net while setting yourself up. Then take him around the showground and preferably into the ring to show him the sights. We then LUNGE (or ride) our horse to have him unstiffen and relax. We do not allow him to tear around or misbehave. He is here with a job to do! Don't grind on him. Work out how long this will likely take the night before so you won't be rushing.

If you see people lunging their horses through the night, with multiple people taking it in turns to wear the horse down before a class at which swabbing is done, you can just about guarantee these people use drugs the rest of the time to keep their horses calm. This should be called out for the abuse that it is, and/or the poor training which goes into these horses, as poor training doesn't equip them to behave properly. The most we have EVER had to use in a very high atmosphere show is ear plugs, on a couple of ponies, just to ensure safety of the child rider. There is no excuse for abuse.

Paying attention to the judge at all times. Brooke Allan with miniature pony Willowmoss Nightingales Song, header Breanna Holmes/Eeson. Photo Rebecca Holmes.

Next, groom or sponge off any dirty patches that may have occurred overnight. Follow up with a soft body brush. Brush the tail thoroughly. Put on the false tail if you use one. Cover up any blemishes as best as you can. Make-up will enhance the horse's appearance, provided you use it sparingly. Paint hooves on top, back, and underneath. Also, put a dab of black on the chestnuts of the horse's legs. Cover your horse with a rug, hood, and tail bag. You can keep his hooves out of the mud or dust by putting down rubber matting, a piece of artificial lawn, or an old carpet.

Learn by watching what others do, but don't dawdle and chat! Don't shout so all and sundry can hear.

Get dressed to look as presentable as your horse. It is important you dress appropriately for the breed. ALWAYS wear a hat (ladies wear a hairnet), gloves, and long sleeves. Put on the bridle/halter, ensure all straps are in their keepers, collect your whip (if required), and pin your number. We like to use a number holder. Take a bag or bucket containing a soft brush, cloth, sponge, water bottle, and fly spray to ringside for last-minute touch-ups. Don't forget your camera!

Led-in class tips

Before the class:

- Work your horse before the class and put him away.

- Enter another class first, e.g. showmanship, before the main class, to calm yourself or your horse and know you are ready for your main class.

- Carry a whip (if allowed) to signal and position your horse.

- Be aware of what is happening around you – the judge, the photographer, the announcer, stewards etc.

The class:

- Be on time. Create a good first impression!

- Walk out – Show overtrack and head nod.

- If the horse in front is dawdling or misbehaving, stop or pass.

The line-up:

- Line up a clear distance from the previous horse.

- Stand your horse in the breed's preferred pose. He must not fidget!

- Stand in formal position and set your horse up.

- Look like you're the winner!

Common mistakes:

- Standing with your back to the judge, or between the judge and your horse.

- Standing at your horse's shoulder.

- Not keeping eye contact with the judge.

- Not focusing on the judge and what your horse is doing.

- Allowing your horse to fidget. (Correct a fidgeting horse without making it obvious.)

- Allowing your horse to swing out or sniff another horse.

- Turning a circle to reposition your horse. Never circle – always straighten up.

- Not correcting a stallion/gelding letting down.

- Not correcting a horse resting a leg or looking bored.

Stay in sync with your horse. Chalani Minerva - Photo by Kangra

The Workout:

- Acknowledge the judge at the beginning and end.

- Listen to instructions and follow them exactly.

- Turn your horse away from you.

- Be forward, straight, and precise to markers.

- Head directly to the judge or straight past, as requested.

- Halt, stand, and pose your horse until dismissed.

- Go behind others if re-entering the line-up.

Final moments:

- It is not over until it's all over. Keep showing your horse.

- Remember, you are all winners until the judge places the ribbon!

- Congratulate the winner! If it is not you, be a gracious loser. Your turn will come.

- Etiquette requires that you don't leave the ring until requested by the steward and after the winner.

- Learn from the experience. Relax and have fun.

- Debrief with your strappers and thank them. Look after them with an icecream or a beer!

Watch and learn from those more successful than you. There is no-one who cannot be beaten with a little hard work. If possible, debrief with the judge or someone who can observe you or take a video. Then, go to the stall and look at the winner's horses. Someone can always step up and take the winner's place, so keep at it. That person can be YOU!

Congratulating the winner. Kate Feeny (winner) with Wishaw Glen Cherish The Moment and Brooke Allan (2nd place) with Argus Park Pandora, owned by Nigel Burton. Photo Rodney's Photography.

Chapter 37

Final Words

The future is big, but the present is just a thin line.

A breeder is usually thinking about the future. Learn to enjoy the moment. Break the cycle of rushing, stressing, or being a robot to your horses, to get more done. Routines bind you in the horse world, so consider whether they are essential. Break the cycle by simply enjoying your horses. When "in the zone," you forget about time and chaos. Only do those things which are important or enjoyable to move you forward. You won't succeed otherwise. Make sure you take time out for yourself from time to time. Wash the dishes, clean tack, groom a horse, do yoga, go for a trail ride, grab a coffee and read a book; whatever takes you out of your head space. Solitude is restorative so nurture yourself. It requires the fortitude to say "no." Saying "no" to people is not unlike setting boundaries for horses. It must be done for the long term benefit.

Each day you may encounter deadlines and "spot fires" to put out, not necessarily of your own making. Don't over-analyze or beat yourself up for something out of your control. Ask yourself, can worse things happen? We can become so embroiled in our emotional stuff that it takes hold. Don't sweat the small stuff. Move on. See it as part of the big picture. Endless details will always beg for attention, time, and energy. If you find life's challenges too big to make time for your horses, perhaps find lease homes for your most valued stock for a few years, your seed stock, so that you can work on one thing at a time, while selling the rest. Speak to your mentor about options.

You may have thought your problem was "time starvation," when in truth, it was in the way you assigned priorities in your decision-making process. Have you allowed the urgent to crowd out the important? One of the greatest skills you can develop is telling the two apart and assigning the correct time to each. You can master this skill with a little determination and patience. Be willing to wait when the results of your actions are not immediately evident and put the urgent aside for the important.

Make the most of opportunities as they present themselves. Watch out for them through research and due diligence. It is no use regretting lost opportunity; learn from it. Another opportunity always comes up. It is also important not to be too risk averse. Fear is a great obstacle. If you are a scaredy cat, you are not likely to take advantage of an opportunity, even one with a better-than-average chance of success. What growth opportunity do you see out there? What is *your* missing piece? How can you change that?

Shangrila Al Don't Buck Me Off, Miniature filly. L-R Jenni Phillips (steward) myself (judge) with owner Mick Sinclair. Photo Shadows Farm Photography.

Take the time to get presentation photos. Sometimes all the stewards want in! Judge Larry Cutler, handler Kim Ide with Chalani Sunstream. Photo by Kangra.

True freedom comes from discipline.

I have often encouraged an owner to have a go at taking their youngster to a bigger event than their comfort zone. They can't believe it if they are successful! If this means getting a trainer to help you, go for it. Follow your dreams. This might be your first (or last) opportunity to try something bigger. You will likely regret the shots you don't take.

"My greatest mistakes were errors of omission, not commission" – Warren Buffett

"You miss 100% of the shots you don't take." – Wayne Gretzy, hockey legend

Do not be fooled into believing horse people are conversant with the facts. Many are simply repeating gossip, old wives' tales, or long-held incorrect beliefs, especially on social media which at best can be described as "horsey entertainment." Ignore all of this to step up above it. Find new friends if necessary.

"Although your own internal measurements are the most important, seek external feedback on your progress toward your goals. When you do, be sure it is from people who are truly interested in seeing you succeed. Don't seek feedback from fair-weather friends, competitive peers, or any person who doesn't have your best interests at heart. Neutral doesn't count. Get feedback from someone who is on your side but will still be objective and honest with you.

"I've observed time and again that misery truly does love company. Jealousy creates some of the most miserable people I know. Surpass the achievements of your particular social crowd or your business colleagues, and look out for the slings and arrows of those who wish you were back where they are. You have to dodge the snide remarks and catty comments. Let them roll right off you. Don't internalize them.

"Only pay attention to feedback from those who have similar goals or who are working actively alongside you to achieve goals of their own. Motives and fears run deep. The sympathetic fair-weather friend who supports you and comforts you when you're down may like you best when you are in just that state: down and dependent". – Denis Waitley

Day's end at Chalani

"The world is not driven by greed; it's driven by envy. I have conquered envy in my own life. I don't envy anybody. I don't give a damn what someone else has. But other people are driven crazy by it." – Charlie Munger

"Envy is everywhere, people are consumed by it. There is no other country in the world that envy and the tall poppy syndrome exist like it does in Australia. In most other countries,

success is celebrated, but in Australia the common thing to do when someone is successful is to tear them down. This trait comes from a place of weakness" – Andrew Barnett

Staying ahead of the competition

Who are your competitors? Great businesses do attract imitators and competitors. Copying is the best form of flattery. If you hear put-downs or malicious gossip, you are being noticed. You are succeeding on some level. Truly good businesses are the ones that can outperform competitors and hopefully have long, continuous growth. This won't happen if you can't attract repeat customers and repeat sales. If not, ask yourself, why not? Sometimes you need to spend more money in advertising, training, or purchases, to make more money.

Be consistent. To stay in business, it is the total sale price of a breeder's stock which counts, not the occasional well-publicised high or record price. Turn everything you have learned into consistency, automate your habits, but be flexible. What seems to be missing in the market? Can you capitalize on that missing piece?

The test of a great breeder is not so much where he starts, but his ability to improve on each generation of his stock through responsible choices. He can then place his horses into the hands of the right people through a realistic awareness of the horses' attributes.

There is no such thing as a success ratio, such as "I have bred x good stock, but y average stock or z poor stock." This is a subjective judgment. Your best one might be sitting in someone's paddock. While that might be disappointing, it is not a failure on your part. Perhaps you can offer incentives, or help to this person, to get the horse "out there." This game's only "fail" is if you are not a *responsible, thinking* breeder.

When you become really interested in something, you can accumulate knowledge faster and better than anyone else. When you're passionately interested in something, you will keep pursuing it even when others have given up. You will be using this knowledge in a competitive environment. When others are satisfied, you will keep enquiring. Ultimately, the only reliable approach is accumulating your knowledge piece by piece by letting your passions be your guide, using a scientific approach, and maintaining intellectual honesty. This will give you a competitive edge.

"Real businesses don't really change by day, by week, by months. It takes years for them to grow. And so, you should expect your results to come in slowly, gradually over a long period of time" – Li Lu

The first name is the one people remember. Do you remember who came second to Winx? Or Frankel? Your competition only stands out because they are first on some measurement scale. The statistics, photos, and magic build until they have surpassed others in the race. Winning is merely the shop window. There can only be one winner on each occasion, but there is plenty of room at the top. Change your philosophy from "beating the competition" to "being the best you can be" and "learn from the competition." Then you will be truly successful in more ways than one.

If you are merely resting on past laurels, your stud has stopped proving itself and is in danger of sliding into anonymity.

Becoming future proof

There are master breeders still out there, successful people with philosophies you can learn from. Don't overlook your breed and discipline magazines. Interviews with successful breeders may provide a little snippet of information that keeps you motivated or propels you forward. Read every sentence of your magazine to stay up-to-date and excited about the future of your breed. If you don't read it, why not? Perhaps you've become too cynical and disillusioned and need to revisit your goals.

Keep an open mind. Be open minded enough to listen to those who are successful and smart enough to distinguish these from those who just like to talk a lot. Even if you don't agree with something, you can learn from it. ("I don't like him, but I like his horses.") Just because someone does something differently from you doesn't necessarily make it wrong!

Keep an open mind

Truly successful people give back. Consider your stewardship. How do you look after the land? Or give back to the next generation? How will you nurture young, upcoming stars? Who do you look up to in the industry? Who in the industry do you respect the most? How would you feel if you were to ring and chat with them about the merits of bloodlines, trainers, or other pertinent information? If this thought frightens you, you are still a tadpole in the industry. (Nothing wrong with that, but it pays to know exactly where you are).

What is your health like? Many elderly people breed horses they can no longer handle or care for adequately. As you get older and less physically able, passing on your lines to responsible people is important. Do you have a handover plan? How will you pass on your horses and your legacy? Do you have stored semen from your now deceased genetics which new breeders can utilise? Do you need to scale back now to ensure you can breed a foal or two in your twilight years? If you have lost the passion, or your health is waning, consider no longer breeding. Let your aged mares age with you. Enjoy watching others succeed with the horses you bred by staying in touch with their owners.

Becoming future proof includes planning for your family to take over or selling your stud as a package so that your bloodlines continue in a way that you choose. You can offer advice, but you must let go of control. The most successful studs over many years are those handed down to family ownership and built further by the younger members of the family with newly infused motivation, inspiration and skills.

If you know better, you do better. Keep educating yourself. Knowledge is power. The longer you're in this business, the more you realize that this is not just a horse business but a *people* business.

Ultimately, nobody else is responsible for your life but you. Nobody else is accountable for your actions but you. Therefore, nobody's expectations for you and opinions about you are as important as your own. So make sure those take precedence in your mind over all others. Draw from the knowledge you have deep inside you already. *Trust yourself.*

Life is a series of choices. The beauty is we get to choose. How we respond to things and our decisions from this day forward are all that matter. The past doesn't define us.

> *"Step away from being a victim of your circumstances and make the decision to step into who you were born to be -"a creator." Dare to dream; to actualise those visions that you have buried deeply; to do whatever it takes, to find clarity on your pathway; to have the courage to reconnect with your self-identity so that you can have the healthiest relationship with yourself."*
> – Jess Keenan, Leading Change Experiences

Look at your notebook and see if where you are now is congruent with your goals. Have you changed your vision or goals after reading these chapters? If so, can you add new goals and thoughts to your notebook? If you work hard but still feel you are getting nowhere, perhaps you are putting your effort and energy in the wrong direction. Refine the intangibles, your life's purpose. What changes can you action *now*?

What reflections can you make as a result of reading "*The Thinking Horse Breeder?*" Are you more committed? A lot of the questions I have asked are "open" questions, that is, there is no correct answer, so I suggest pondering them seriously by going back and re-reading chapters to formulate your thoughts, and add them to your notebook. What rings true for you?

By working through this book you have all the insights I can pack into one volume. I have bared my heart and soul. Horse people, being horse people, will disagree with me over some of the things I have said, and have their own way of doing it. Alway think "is there a better way for our horses?" We are all still learning, especially with new research and technologies to assist us. The future is bright for those who dare to dream and put in the work.

It took me 50+ years to learn all this, and I am still learning. I have made truly wonderful friends and am so grateful for being part of the industry. There is no better feeling than taking a homebred horse to the winner's circle and seeing others do it with horses you've bred. You can do this too. Enjoy your journey.

Joseph's Dream Appaloosa Stud mares and foals coming in at daybreak, courtesy Annika Funke-Barnard of Joseph's Dream Stud and Retreat, Namibia. Take a look at their website. It is truly beautiful.□
www.josephsdreamstud.com
□

Some facts and figures

Chalani Australian Stock Horses

290 foals bred

Average = 5 foals bred per year

Injuries resulting in horse's death (includes euthanasia) = 4

Injuries resulting in retirement to broodmare band = 3

Unexplained deaths (horse found deceased) 3

Colic = 7 (2 of these resulting in death)

Stillborn foals = 6

Twins 3x sets = one of each survived

Premature foals resulting in vet treatment = Nil

Dummy foal = 1

Foals with limb deformities requiring intervention = 4

Orphan foals = 2

Foals with hernias requiring surgery = 4

Mares requiring assistance of vet for foaling = 5

Mare deaths due to foaling = 3

Retained afterbirth requiring vet assistance – (<20)

Cryptorchidism = 3

Chalani stud frequencies at the time of writing – your figures and ratios will be completely different.

Some of our success is management, but we have certainly had a large dose of luck. It will help you to know how you are travelling by working out your own ratios every few years.

The horseman

Epilogue

"There are horse trainers, horse traders and horse whisperers. There are show men, show boaters and show-offs. There are fast talkers and would be magicians.

But then there are true Horsemen and Horsewomen, and these are harder to find and sometimes even harder to recognize because they are often tucked away in quiet hidden places, working slowly and silently without national recognition or appreciation.

"Often times, the true horseman or woman does not have the most horses in training or those horses that are exceptionally bred or high priced. Often times, the true Horsemen and Women do not have access to big money owners or run through dozens and dozens of prospects in order to find the few that can take the pressure of aged event prize money or high-profile exhibition. Many times, the true horsemen and women are slow and steady, methodical and patient, training on an individual horse's timeline and not to a rigid show schedule set by the seasons or show management.

"These people recognize a horse's physical and mental capabilities and showcase their assets without sacrificing their bodies or minds. Horsemen and women take their time developing their horses' skills and confidence through traditional steps, one before the next, placing just as much credence in their teaching relationship and equine partnership as they do in show pen results.

"Horsemen and women are humble because their reward comes from within; from knowing that they have taught through kindness, patience, fortitude, and logic. Their rewards coming from creating a confident horse that works with them and not for them, horses that are not scared or intimidated, horses with solid foundations that last season after season and that carry a gamut of riders from the experienced non pro to the Amateur to the Green Reiner. Always Dedicated. Always Patient. Always Consistent.

"Whether it be riding young horses, resurrecting older horses, or maintaining the Steady Eddy, a True Horseman is one of the first ones to throw a leg over in the morning and one of the last ones to pull their boots off in the evening.

"Horsemen and women are a pleasure to watch in the arena or on the ranch as they diligently and patiently impart their knowledge and logic to both horses and students.

"In an era where the horse industry is so economically driven and success is measured primarily in prize money and accolades, the tradition of the true horseman and the process of training horses seems to be changing; giving way to an assembly-line mentality where immediate success and financial compensation take precedence over handcrafted quality and longevity. Dedicated to their craft, loyal to their students, ambitious, hardworking and a role model for anyone interested in making their way in an industry dominated by pressure to build great animals in less and less time, old fashioned horsemen or women are now Artisans, assets to our heritage and traditions and harder and harder to find.

"A thoughtful teacher, a thorough instructor, a gentle hand, a firm guide, a rational yet fearless showman, the greatest compliment that I think could ever be given to someone who works with horses, is to be thought of as a Horseman."

– By Becky Hanson

***

In loving memory of Barry Elliott, a true horseman of the old tradition, died February 2023.

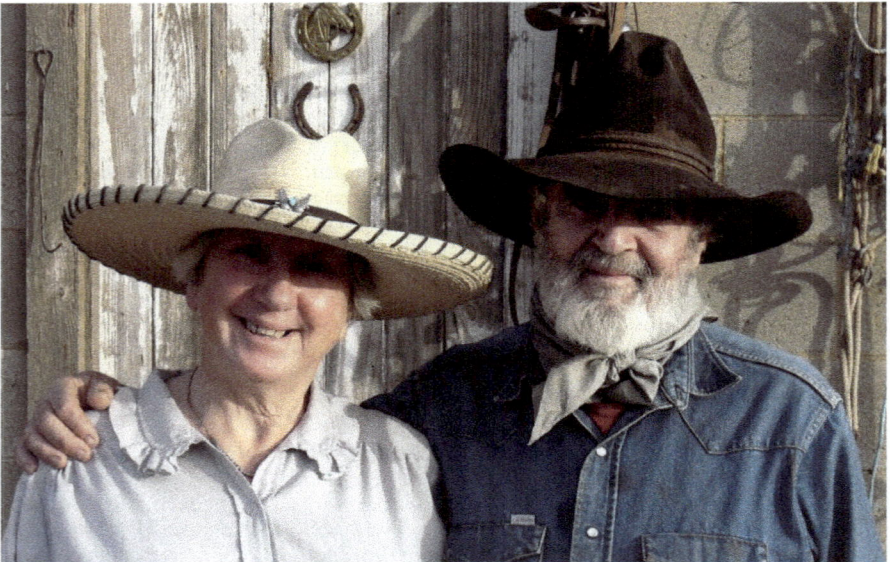

Merrie and Barry Elliott - Photo Claus Thrun

Contact the author

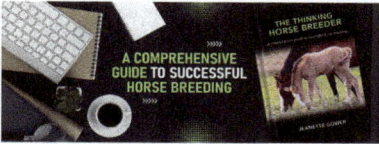

If you enjoyed reading this book, I would be really grateful if you could put a **Review** on Amazon. Just search Amazon - books - *The Thinking Horse Breeder* and follow the prompts. To buy *signed* copies, or for answers to general queries about horse breeding, I can be contacted via *The Thinking Horse Breeder* Facebook page or https://thinkingh orsebreeder.chalani.net/

My contact email is: jeanette.gower@gmail.com Send me an email to let me know one thing you learned from this book. I would love to hear from you.

Chalani can be contacted by email chalani@chalani.net or our Facebook page. Chalani Sunstream HSH is available by frozen semen in USA, and Australia. Photos and video on the *Chalani* Facebook page, the *Chalani* website or You-tube channel.

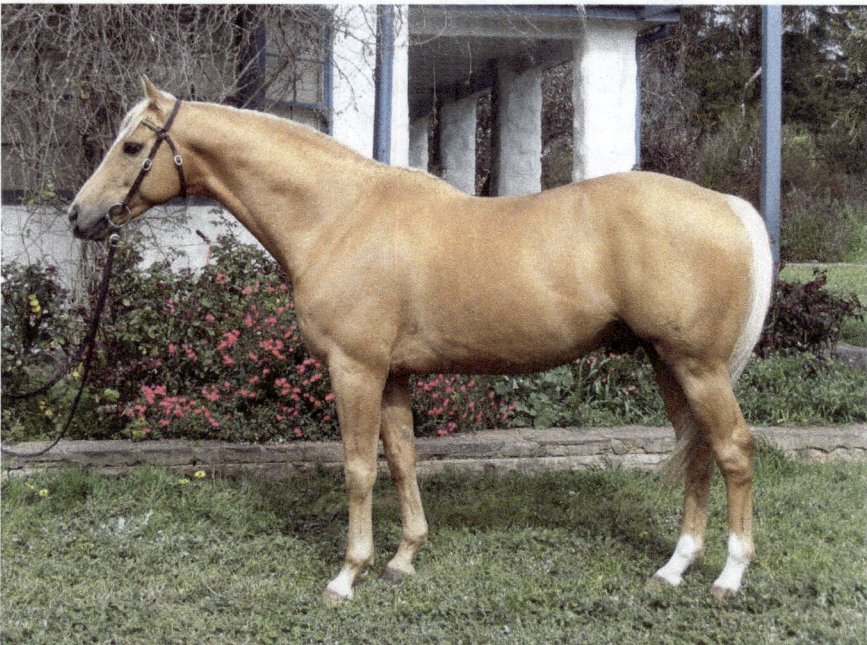

Chalani Sunstream

Chalani Sunstream HSH - with Kim Ide. Photos by Nicole Ceary, Kerri Afford, Horizons, Beauty and the Huntress, Kate Merritt.

Testimonials

"As one of the fortunates chosen to have a read-through of Jeanette's book, I can honestly say I'd recommend it to anyone looking to start out in horse breeding. I believe it will become an invaluable tool to the newbies out there. However, I personally also will find the knowledge imparted so clearly in this book, incredibly useful for my own continuing journey as a horse breeder. I believe even experienced breeders will enjoy looking back over their own breeding journey and seeing the similarities or even things they would've done differently with a little more knowledge and the beauty of hindsight.

"In particular, I found the financial section eye opening. If only I'd had that advice before me when I first started out! For those of us not blessed in this area, it is a priceless take on how to start up and not flounder in a sea of debt and unnecessary expenses but instead, begin to break even or possibly make a profit over many years. In short, if you really are passionate about breeding a beautiful, saleable, well-handled animal, this book is for you. Make the most of the decades of experience so willingly shared with us and dive in. I guarantee you will enjoy the read." *- Kelsey Stafford, Tarrawonga ASHs.*

"Jeanette Gower has been a well-known and respected breeder in South Australia for several decades. She has been a regular contributor to magazines and lectured on various horse husbandry topics for as long as I remember, attending her lectures myself as a teenager. Her first book, *Horse Colour Explained*, was the first book on equine colour genetics published in Australia that brought the topic in a simple and easy to understand format to the average horse owner. During the book's research stages in early days of email, I remember assisting Jeanette in her correspondence with genetics experts overseas, and it was fascinating to be a part of those exchanges.

"Now, she has created a concise and informative guide, using her 50 years' experience as a successful horse breeder to present a thought-provoking information source for newer breeders wanting to plan for the future. The book is not a "how to" guide on horse breeding or handling, but rather presents an overview of all the topics and scenarios that a breeder should consider when setting up or expanding a horse stud with longevity in mind. It allows a new breeder the freedom to do things their own way, while encouraging them to think, plan and to make informed decisions on how to create their own successful breeding operation." *– Katherine Szalay Evans, Halado Park Lipizzaners.*

***

Stop press: Oct 2023

Rannock (see story in Introduction) was awarded one of the highest honours of the Australian Stock Horse Society, by being inducted into the ASH Hall of Fame, for his contribution to the breed.

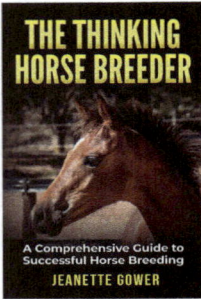

About *The Thinking Horse Breeder*

Horse breeding is more complex than you think. Set yourself up to succeed in your passion for horses. How would you go about establishing a horse stud? Are you breeding horses and want to take it to the next level? Do you want to make better choices, breed better horses, improve your results? It can be a lonely journey without someone to learn from, to show you how to lessen the mistakes along the way, and to steer you in the right direction. The Thinking Horse Breeder is a step by step, treasure trove of practical information for those who want to thrive and succeed doing what they love.

This book discusses selection of foundation stock, planning matings, breeding methods, genetic diseases, conformation and temperament, foaling down, raising foals, young horse training, common problems, financials, promotion, photography and ethical considerations, in an easy-to-read, authentic style. Each chapter can be read and re-read for new insights. It will challenge your thinking and give you the art, science and tools for success. Everything you need to know is here in a simple, non-technical format based on the author's experiences and reflections over 50+ years. For aspiring, hobby or serious, established breeders, this will be an invaluable guide to be read over and over, so you too can master the inevitable challenges and be successful.

Also by Jeanette Gower – The Rannock Legacy

In October 2023, Rannock was inducted to the ASH Society's Hall of Fame for his contribution to the breed as a foundation sire. Time is now full circle to tell his story.

"*The Rannock Legacy*" is not just a book—it's a journey through time, tracing Rannock's lineage from his humble beginnings in the Hunter Valley to his legendary status as a foundation sire in South Australia. Written with meticulous research and heartfelt passion, this book offers a comprehensive exploration of Rannock's life, lineage, and lasting impact on the Australian Stock Horse community.

The rich history of the Australian Stock Horse comes to life through the captivating story of Rannock and his enduring legacy. What sets "*The Rannock Legacy*" apart is its intimate portrayal of Rannock and his descendants through personal stories, rare photographs, and firsthand accounts from breeders and owners. Discover the resilience, athleticism, and generosity of these remarkable horses as they carve their place in equine history, leaving an indelible mark on the breed for generations to come.

Whether you're a seasoned breeder, an equine enthusiast, or simply curious about the rich heritage of Australian Stock Horses, "*The Rannock Legacy*" is a must-have addition to your library.

Order your copy and experience the magic of "The Rannock Legacy" or go to https://thinkinghorsebreeder.chalani.net/index.php/the-rannock-legacy/ Also available on Amazon and Books.by. https://books.by/jeanette-gower

Also by Jeanette Gower

A straight-talking guide to smarter buying

Buying a horse is exciting—but it can also be a minefield.The wrong choice can cost you time, money, and heartache. Whether you're looking for a safe pleasure mount, a competition partner, or a breeding prospect, this book gives you practical, no-nonsense advice, and the checklists you need to make the right decision.

Inside, you'll discover:

Essential preparation tools to avoid impulse purchases and choose the horse that best meets your goals, whether as a competitor, leisure rider, a young horse buyer,or a breeder.

- Step-by-step guidance and checklists for evaluating horses—from uncovering red flags, to assessing conformation and temperament like a pro.

- Expert strategies for navigating buying with confidence, including must-ask questions,legal tips, and negotiation methods.·

- Buying at auction, and other methods of purchase, with real-life success stories.

- Post-purchase advice to seamlessly integrate your new horse into your care.

You don't need decades of experience to buy the right horse. This guide gives you the tools to make informed, confident decisions—while avoiding costly mistakes. *Buy The Right Horse* is the ultimate companion for finding your dream horse—with zero regrets!

Available from:

Books.by/jeanette-gower or go to my Author page: https://thinkinghorsebreeder.chalani.net/or Amazon.

REFLECTIONS ON 50+YEARS OF HORSE BREEDING

Tired of theoretical advice & generic tips? Ask questions. Get insights, practical advice, resources, & inspiration. Master the art, science & tools for excellence from a seasoned horse breeder who's been in the trenches.

Launched 2 years ago

Join me on Substack:

https://jeanettegower.su

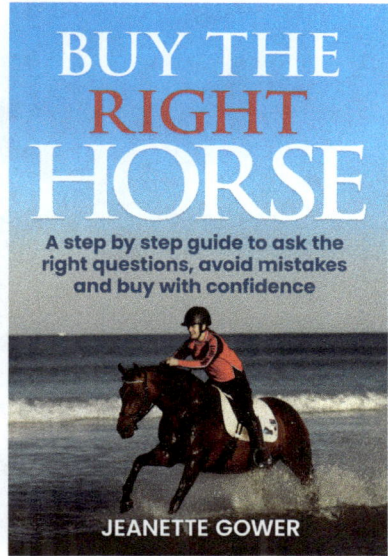

Type your email...   Subscribe

www.ingramcontent.com/pod-product-compliance
Lightning Source LLC
Chambersburg PA
CBHW071552210326
41597CB00019B/3213